T0223019

Earth Summit II

Earth Summit II
Outcomes and Analysis

by Derek Osborn and Tom Bigg

earthscan
from Routledge

Routledge
Taylor & Francis Group

LONDON AND NEW YORK

First published in the UK in 1998 by
Earthscan Publications Limited

This edition published 2013 by Earthscan

For a full list of publications please contact:

Earthscan
2 Park Square, Milton Park, Abingdon, Oxon OX14 4RN
Simultaneously published in the USA and Canada by Earthscan
605 Third Avenue, New York, NY 10017

Earthscan is an imprint of the Taylor & Francis Group, an informa business

A catalogue record for this book is available from the British Library

ISBN 13: 978-1-85383-533-9 (pbk)

Typesetting and page design by PCS Mapping & DTP, Newcastle upon Tyne
Cover design by Declan Buckley

Contents

List of Acronyms

CSD	Commission on Sustainable Development
DAC	Development Assistance Committee
DPCSD	Department of Policy Coordination and Sustainable Development
DREAMS	Development Reconciling Environmental And Material Success
ECOSOC	United Nations Economic and Social Council
EU	European Union
GEF	Global Environment Facility
GLASOD	Global Assessment of Human Induced Soil Degradation
G77	The Group of Developing Countries
IACSD	Inter-agency Committee on Sustainable Development
IAEA	International Atomic Energy Agency
IFIs	International Financial Institutions
ICPD	International Conference on Population and Development
IPCC	Intergovernmental Panel on Climate Change
IWGCSD	Intersessional Working Group of the Commission on Sustainable Development
MAI	Multilateral Agreement on Investments
NGO	non-governmental organization
ODA	Official Development Assistance
OECD	Organization for Economic Cooperation and Development
OPEC	Organization of Petroleum Exporting Countries
POP	persistent organic pollutant
SLUDGE	Slightly Less Unsustainable Development Genuflecting to the Environment
UNCED	United Nations Conference on Environment and Development
UNCTAD	United Nations Conference on Trade and Development
UNDP	United Nations Development Programme
UNEP	United Nations Environment Programme

UNESCO	United Nations Educational, Scientific and Cultural Organization
UNGASS	United Nations General Assembly Special Session
UNIDO	United Nations Industrial Development Organization
WBCSD	World Business Council on Sustainable Development
WSSD	World Summit for Social Development
WTO	World Trade Organization

Foreword

One of my first international engagements as Prime Minister was to attend the UN Special Session in New York last June. We would have liked more to have been achieved there. Nevertheless, the international community did renew its commitment to sustainable development and agree a practicable work programme for the next five years. Important initiatives were taken on freshwater, forests, oceans and energy. There was also agreement that the UN's Commission on Sustainable Development should have a strategic role in taking these initiatives forward. These are solid and worthwhile achievements.

I am proud of the role that the UK played at the Special Session. The strength and depth of our team, including many representatives from outside government, demonstrated our commitment to sustainable development. It also showed that effective action means bringing the domestic and international agendas together, something I know that UNED-UK and other NGOs in this country have been working hard to promote.

As we take up the Presidency of the European Union next year and the Chair of the G8, we will continue to build up our partnership with those outside government. We must create a consensus on the need for sustainable development. I am sure that this review of the Special Session will help develop that consensus.

The Right Honourable Tony Blair MP,
Prime Minister of the United Kingdom of Great Britain
and Northern Ireland
November 1997

Foreword

The Nineteenth Special Session of the United Nations General Assembly was the first milestone, and an important benchmark, in national and international implementation of Agenda 21. Taking place five years after the United Nations Conference on Environment and Development (UNCED), it provided an opportunity to take stock of what had – and what had not – taken place, to review commitments made in Rio de Janeiro in 1992, and, accordingly, to map out a programme of further work including the programme for the United Nations Commission on Sustainable Development (CSD) to the year 2002.

The Special Session did not result in any major breakthroughs. It did not fully meet the expectations of developing countries which had hoped for new commitments and initiatives from developed countries, particularly regarding the provision of new and additional financial resources; nor of those countries which expected that the Session would agree on specific timebound quantitative environmental targets, particularly for the reduction of greenhouse gas emissions.

However, it did both successfully carry out a frank assessment of progress and adopt a *Programme for the Further Implementation of Agenda 21* to accelerate progress towards sustainable development. Among the more salient accomplishments were the following:

1. a strong reconfirmation of a political commitment from all members of the international community, as well as from all major groups of civil society, to sustainable development and of the key role to be played in this area by the United Nations;
2. the clarification of the specific roles of various organs and institutions in further work on environment and sustainable development;
3. a more focused Programme of Work for the CSD and a clear and continuing commitment to a Commission that is open, transparent and participatory;
4. reconfirmation of the agreed UNCED targets and commitments for Official Development Assistance (ODA), and a call for intensified efforts to reverse the declining ODA/GNP ratio;

5. continuation of a political process under the auspices of the CSD on forests and a more focused consideration of the modalities for a possible legally-binding instrument in this area;
6. the beginning of an intergovernmental process within the CSD on both freshwater and on energy;
7. a better understanding of the importance of a stronger commitment to such issues as tourism, transport, information, and changing production and consumption patterns; and
8. a number of new practical agreements in specific areas, such as the one on a world-wide phase out of lead from gasoline.

The UN Conference on Environment and Development brought political visibility to the issues of sustainable development and energized the participation of both public and civil society. Five years later, we understand that we are in for the long haul. What the Special Session of the General Assembly clearly demonstrated was: that the commitment to Agenda 21 endures; that significant institutional changes have occurred at national, regional and international levels to respond to Agenda 21; and that progress continues. The agenda of the Commission for the next five years is a challenging and exciting one. The United Nations and all of its many partners look forward to meeting that challenge.

I believe that the contents of this book will assist the international community in this effort. Its authors played key roles in the process leading up to and concluding with the Special Session of the General Assembly, and have had the opportunity to reflect on the most important lessons to be learnt from this exercise. They are thus well placed to propose a number of interesting ideas on how to ensure successful preparations for the ten year review that will take place in the year 2002.

It was my pleasure to have read this book, and it is my pleasure to recommend it to all those who are interested in the issues of sustainable development and multilateral diplomacy.

Nitin Desai
United Nations Under-Secretary-General
for Economic and Social Affairs
February 1998

Addresses of Key Organizations

Basel Convention Secretariat,
Geneva Executive Center,
15 Chemin des Anemones,
Building D,
1219 Chatelaine Geneva
Switzerland
Tel: 41 22 979 9111
Fax: 41 22 797 3454

Centre for Sustainable Communities
University of Washington
Seattle
Washington
USA
Tel: 1 206 616 2035
Fax: 1 206 543 2463
Email: lawrejg.washington.edu

Climate Change Secretariat
Geneva Executive Centre
11–13 Chemin des Anemones
1219 Chatelaine
Geneva
Switzerland
Tel: 41 22 979 9111
Fax: 41 22 979 9034
Email: secretariat, unfccc@unep.ch
Web site: http://ww.unep.ch/unfc-
  cc/html

Consumers International
24 Highbury Crescent
London N5 1RX
Tel: 44 171 226 6663
Fax: 44 171 354 0607
Email: prodec@consint.dircon.co.uk

Convention on Biological Diversity
Secretariat
World Trade Centre
413 St Jaques Street, Office 630
Montreal,
Quebec H2Y 1N9
Canada
Tel: 1 514 288 2220
Fax: 1 514 228 6588
Web site http://www.unep.ch/biodiv

Convention to Combat
Desertification
11/13 Chemin des Anemones,
BP 76, 1219 Chatelaine,
Geneva
Switzerland
Tel: 41 22 979 9411
Fax: 41 22 979 9030
Email: secretariat.incd@unep.ch

CSD NGO/MAJOR GROUPS
    STEERING COMMITTEE
Southern Co-Chair:
Esmerelda Brown
Southern Diaspora Research Centre,
391 Eastern Parkway,
New York NY 11216 USA
Tel: 1 212 682 3633 (work)
Fax: 1 212 682 5354
Email: umcgbgm@undp.org

Northern Co-facilitators:
Michael McCoy
Citizens Network
77 Perry Street
Apartment 1B

New York NY10012
USA
Tel: 1 212 243 1855
Felix Dodds (see UNED-UK)
Web site
   http://www.igc.apc.org/habitat/csd
   -97

Development Alternatives
B-32 Tara Crescent,
Qutab Institutional Area,
New Delhi 110016
India
Tel: 91 11 66 5370
Fax: 91 11 68 66 031

Earth Council
Apdo 2323
San Jose
Costa Rica
Tel: 506 223 3418/256 1611
Fax: 506 225 2197
Email: eci@terra.ecouncil.ac.cr

Earth Negotiations Bulletin
c/o IISD
161 Portage Avenue East
6th Floor
Winnipeg
Manitoba R3B 0Y4
Canada
Tel: 1 204 958 7710
Fax: 1 204 958 7710
Email: enb@econet.apc.org
Web site http://www.iisd.ca/linkages

Earthscan Publications Limited
120 Pentonville Road
London N1 9JN
Tel: 44 171 278 0433
Fax: 44 171 278 1142
Email: earthinfo@earthscan.co.uk
Web site http://www.earthscan.co.uk

Environmental Development Action
   in the Third World
ENDA Maghreb
196 Quartiert OLM
Rabat-Souissi
Morocco
Tel: 212 775 6414
Fax: 212 775 6413
Email: magdi@endamag.gn.apc.org

Environmental Liaison Centre
   International (ELCI)
PO Box 72461
Nairobi
Kenya
Tel: 254 256 2015
Fax: 254 256 2175
Email: elci@elci.sasa.unep.no

Food and Agriculture Organization
   of the UN
Via delle Terme di Caracalla 00100
Rome
Italy
Tel: 39 6 57971
Fax: 39 6 578 2610

FIELD
46 Russell Square
London WC1
Tel: 44 171 637 7950
Fax: 44 171 637 7951

Friends of the Earth
26–28 Underwood Street
London N1 7JQ
Tel: 44 171 490 1555
Fax: 44 171 490 0881

Global Environmental Facility
1818 H Street NW
Washington, DC 20433
USA
Tel: 1 202 473 5102
Fax: 1 202 522 3240
Email: anpraag@worldbank.org
Web site http://www.oneworld.org/
   panos

HABITAT
UN Centre for Human Settlements
United Nations Office at Nairobi
PO Box 30030
Nairobi
Kenya
Tel: 254 262 4260
Fax: 254 262 1234
Web site
    http://www.igc.apc.org/habitat

International Confederation of Free
    Trade Unions (ICFTU)
Boulevard Emile Jaqmain 155 B1
1210 Bruxelles
Belgium
Tel: 32 2224 0211
Fax: 32 2201 5815
Email: ICFTU@GEO2.poptel.org.utc

International Council for Local
    Environmental Initiatives (ICLEI)
8th Floor
Toronto
Ontarion M5H 242
Canada
Tel: 1 416 392 1462
Fax: 1 416 392 1478

International Institute for
    Environment and Development
3 Endsleigh Street
London, Wc1H 0DD
Tel: 44 171 388 2412
Fax: 44 171 388 2826

International Labour Organization
    (ILO)
4 routes des Morillons
CH-1211 Geneva 22
Swtizerland
Tel: 41 22 799 6111
Fax: 41 22 798 8685

International Chamber of
    Commerce
38 Cours Albert 1er
75008 Paris
France
Tel: 331 4953 2926
Fax: 331 4953 2859

International Monetary Fund
700 19th Street NW
Washington, DC 20431
USA
Tel: 1 202 623 7000
Fax: 1 202 623 4661

OECD
2 rue Andre Pascal
75775 Paris Cedex 16
France
Tel: 33 1 45 248200
Fax: 33 1 49 104276
Web site http://www.oecd.org

Oxfam
274 Banbury Road
Oxford OX1 7DX
Tel: 44 1865 312 389
Fax: 44 1865 312 417

Ozone Secretariat
UNEP
PO Box 30552
Nairobi
Kenya
Tel: 254 2 521 928
Fax: 254 2 521 930
Email: madhava.sarma@unep.no
Web site
    http://www.unep.org/unep/secre-
    tar/ozon

Third World Network
228 Macalister Road
10400 Penang
Malaysia
Tel: 60 4 226 159
Fax: 60 4 226 4505
Email: twn@igc.apc.org

UN Department for Policy
    Coordination and Sustainable
    Development (DPCSD)
New York
NY 10017
USA
Major Groups Focal Point: Zehra
    Aydin
Tel: 1 212 963 8811
Fax: 1 212 963 1267
Email: aydin@un.org
Web site http://www.un.org/dpcsd

United Nations Development
    Programme
1 United Nations Plaza
New York
NY 10017
USA
Tel: 1 212 906 5000
Web site http://www.undp.org

United Nations Environment and
    Development UK Committee
c/o UNA
3 Whitehall Court
London SW1A 2EL
Tel: 44 171 839 1784
Fax: 44 171 930 5893
Email: una@mcrl.poptel.org.uk
Web site
    http:/www.oneworld.org/uned-uk

UN ECOSOC NGO Unit
Room DC-2 2340
United Nations
New York
NY 10017
USA
Tel: 1 212 963 4842/3
Fax: 1 212 963 4968

United Nations Environment
    Programme
Information and Public Affairs
PO Box 30552
Nairobi
Kenya
Tel: 254 262 3292
Fax: 254 262 3927/3692
Email: ipaunep@gn.apc.org
Web site http://www.unep.no

UN NGO Committe on the
    International Decade for the
    World's Indigenous Peoples
109 West 28th Street
New York
NY 10001
Tel: 1 212 564 3329

UNEP Publications Distribution
SMI Limited
PO Box 119
Stevenage
Herts SG1 4TP
Tel: 44 1438 748 111
Fax: 44 1438 748 844

UNEP Industry Office
Tour Mirabeau
39–43 quai André Citron
75739 Paris
Cedex 15
France
Fax: 33 1 4437 1474

UN Non-Governmental Liaison
  Service (NGLS)
Room 6015
866 UN Plaza
New York
NY 10017
USA
Tel: 1 212 963 3125
Fax: 1 212 963 3062

UNGLS
Palais de Nations
1211 Geneva 10
Switzerland
Tel: 41 22 798 5850
Fax: 41 22 907 0057

World Bank
1818 H Street NW
Washington, DC 20433
USA
Tel: 1 202 477 1234
Web site http://www.worldbank.org

World Business Council for
  Sustainable Development
160 route de Florisant
1231 Conches
Geneva
Tel: 41 22 839 3100
Fax: 41 22 839 3131
Email: wbcsd@iprolink.ch

Women's Environment and
  Development Organization
  (WEDO)
355 Lexington Avenue
3rd Floor
New York
NY 10017-6603
USA
Tel: 1 212 973 0325
Fax: 1 212 973 0335
Email: wedo@igc.apc.org
Web site http://www.wedo.org

World Health Organization
1211 Geneva 27
Switzerland
Tel: 41 22 791 2111
Fax: 41 22 791 0746

World Trade Organization
Centre William Rappard
154 rue de Lausanne
1211 Geneva 21
Switzerland
Tel: 41 22 739 5111
Fax: 41 22 731 4206
Web site
  http://www.unic.org.wto/welcome
  .html

Worldwide Fund For Nature
Avenue du Mont Blanc
CH-1196 Gland
Switzerland
Tel: 41 22 364 9111
Fax: 41 22364 4238

1

A Tale of Two Cities:
From Rio to New York

The United Nations Conference on Environment and Development (UNCED) was held in Rio de Janeiro in June 1992. Popularly known as the Earth Summit, it was attended by over a hundred heads of state and government, more than had ever attended an international conference before or since, indicating the importance that the world attached to the occasion. It agreed an ambitious programme (Agenda 21) for promoting sustainable development throughout the world. Conventions on climate change and biological diversity and a declaration of 27 principles for sustainable development were also agreed.

Five years after this momentous conference the United Nations decided to take stock at a United Nations General Assembly Special Session (UNGASS) in New York in June 1997. Again, many world leaders were in attendance at what has been variously known as Earth Summit II, Earth Summit + 5, UNGASS or occasionally by its formal title – the 19th Special Session of the General Assembly to review progress achieved in the implementation of Agenda 21. This commentary uses the term Earth Summit II throughout.

The purpose of this book is to assist in the understanding of Earth Summit II by drawing together in a single place some of the key documents, and providing a brief commentary on the main texts. The object of the commentary is to set out what some of the main parties hoped to achieve in New York, how far they were successful, what further action is now in hand and what conclusions can be drawn for the future.

The Earth Summit in 1992 was a time of fears and hopes: fears about the deterioration of the global environment and the pressures that humankind is placing on it; hopes that the nations of the world gathered together could face up to these problems and set human development on a new and more sustainable path.

It was the best of times. It was the worst of times. Best, because so much was attempted – a comprehensive overview of all the development needs and pressures on the environment throughout the world and the mapping out of a comprehensive programme of action in Agenda 21 to address them. Worst, because the programme was weakened at the outset by ambiguities and evasions in the text, and by half-hearted commitment from many of the participating countries who adopted it.

Some progress has been made since Rio, but only in particular areas and on particular subjects. The Commission on Sustainable Development (CSD) was established by the United Nations after Rio in order to help maintain and monitor progress on Agenda 21. It has developed a system of national reporting on implementation and has provided a useful forum for ministers and officials to review progress annually. There has been a gradual extension and strengthening of international environmental agreements. At the national level some aspects of the environment have been improved in some countries. At the local level there has been an explosion of community activity throughout the world under the banner of Local Agenda 21 and similar programmes helping to make sustainability more of a reality.

But not enough has been done yet; and on many of the major global issues the position is still deteriorating. Development has been rapid in some countries but there has been stagnation in others. Poverty and inequality are spreading in many parts of the world. Fresh water supplies are dwindling through overuse and pollution. Greenhouse gases are accumulating and the threat of damaging climate change is growing. Forest cover continues to shrink. The oceans are overfished and stocks are shrinking. Much development has been uneven, inequitable and unsustainable.

Sadly, public attention and the collective political will throughout the world to tackle these issues constructively, creatively and cooperatively also seems to have diminished since Rio. Many countries seem more concerned to deplore the failings of others than to deal adequately with their own problems in achieving sustainability.

Crucially, the global political deal that was struck at Rio has come unstuck. At Rio the countries of the North agreed to make new and additional resources available to the South to enable them to handle their development in a more sustainable and environmentally friendly way. But, with some honourable exceptions, the North has not delivered on this promise and the total of official aid has instead shrunk by 20 per cent over the past five years. The South, not surprisingly, feels badly let down by this, and even less inclined to commit itself to any difficult policy measures to make its development more sustainable.

The countries coming together again in New York in 1997 did not want to rewrite Agenda 21. There was almost universal agreement that there was nothing wrong with Agenda 21 as a programme. The trouble was that it had not been implemented vigorously enough. What was needed was some means of breathing new life and action into the programme. One characterization (made in New York) of the five years since Rio, was that they have produced little more than 'Slightly Less Unsustainable Development Genuflecting to the Environment' or SLUDGE. It was urged at the beginning of the review meeting in New York that what the world needs now is to move from SLUDGE to 'Development Reconciling Environmental And Material Success' or DREAMS.

The review process in New York in 1997 tried hard to face up to the reality of all this. It explored all the different sectors and took stock of progress. It faced up to the growing problems of poverty, inequality and environmental degradation in many parts of the world. It identified some areas where promising actions already under way could be reinforced, and other areas where new action is needed. It agreed priorities and a programme of work for the CSD for the next five years.

It cannot truly be claimed, however, that Earth Summit II succeeded in generating enough political attention and momentum to make for real movement on some of the key issues. Nor did it bring about a new spirit of international cooperation on these matters. It produced workmanlike and serviceable conclusions, but no political inspiration. There was a gulf between the inspirational words of many of the world leaders who addressed the Special Session and the more prosaic document that was agreed in their name – a gulf between rhetoric and reality.

One main conclusion for all those concerned with these matters is that the next Earth Summit in 2002, ten years after Rio, will need to be much better prepared if progress is to be made. Above all there will need to be a renewed effort around the world to focus attention on the issues and to build political consciousness and determination to achieve real results. DREAMS cannot be realized without a sustained political effort at all levels of society and in all parts of the world. Nothing will be achieved through waiting, like Mr Micawber, for something to turn up.

2

Hard Times: Earth Summit II

THE PROCESS

The five-year review of progress after Rio was already envisaged and planned for at Rio in Agenda 21.[1] Detailed arrangements for it were planned by the United Nations during 1996 and finalized by the General Assembly in December 1996 (Resolution 51/181 of 20 January, reproduced on page 69). The central element of the review was planned from the outset to be a week-long Special Session of the General Assembly from 23 to 27 June 1997. It was hoped that this session would be attended by heads of state and government, and that their presence would ensure thorough preparation of the occasion and broadly-based commitment to the process by countries and all the relevant departments of government.

The December Resolution provided that the Intersessional Working Group of the Commission on Sustainable Development (IWGCSD) from 24 February to 7 March 1997 should be devoted to preparing for this Special Session, and that the 1997 meeting of the CSD itself (its fifth annual session) from 7 to 25 April should also be devoted to negotiating the possible conclusions and preparing for the Special Session. In the event, more time was found to be necessary, and an additional week of negotiations and preparations took place in New York from 16 to 21 June (the week before the Special Session) in order to bring all the discussions and documents to a suitable state for high-level participation by heads of governments, ministers and others during the week of the Special Session itself.

In preparation for this series of meetings the Secretariat of the CSD produced a set of reports analysing in detail progress since Rio on each of the Chapters of Agenda 21, together with a summary overview report (reproduced on page 78). This is an extremely useful series of reports, which assembled a great deal of information about national and international activity on sustainable development, and provided the basic background for all the 1997 negotiations.

THE INTERSESSIONAL

At the start of the intersessional meeting in February Ambassador Celso Amorim of Brazil and Mr Derek Osborn of the United Kingdom were elected as co-chairs of the meeting. They were invited to conduct the meeting with a view to producing a co-chairs' text by the end of the two-week session, which could then serve as the basis for negotiation at the meeting of the CSD.

In the first week the co-chairs invited all official delegations and representatives of the major groups: women, youth, indigenous people, non-governmental organizations, local authorities, trade unions, industry, the scientific community and farmers to submit ideas and proposals for the forthcoming negotiation. They asked for maximum creativity and receptivity at this stage – there would be time for weeding out weaker proposals later. At the end of the first week the co-chairs produced a first text incorporating as many as possible of these creative ideas and proposals. This was then subject to debate during the second week at the end of which the co-chairs produced a further draft text for submission to the CSD.

The debate covered all the wide range of issues that are included in Agenda 21. On all of them the need for further action was noted and ideas for strengthening implementation were identified. Some countries were unwilling to go beyond this in placing emphasis on particular subjects, but most agreed that six areas in particular emerged as deserving special effort and attention:

- The need to combat poverty and growing inequality in the world, and particularly to bring help to the poorest countries in the South, which have lost out on recent economic growth elsewhere, and whose very poverty exacerbates and is exacerbated by their environmental problems such as tree loss, drought, desertification, soil loss, etc.
- Associated with this the need to arrest the decline in levels of official development assistance, or to find other means of bringing effective help to the poorest countries.

- The need to bring fresh water and sanitation to the hundreds of millions of people who lack them at present, at the same time as dealing with the long-term problems of dwindling water resources and increasing pollution of water in many parts of the world.
- The need for a clear global strategy to deal with the climate change issue (together with the related transport and energy issues).
- The need to establish an effective ongoing process to promote the sustainable management of forests throughout the world.
- The need for more effective international cooperation and political impetus to protect the marine environment and halt the catastrophic decline of fish stocks in many parts of the world resulting from competitive over-fishing.

There were many other significant issues on the agenda as well, but it was widely felt that it would be on these six that the main attention of the world would focus and on which the success or failure of the session would depend.

Rio + 5

From 13 to 19 March 1997 a highly ambitious forum was held in Rio de Janeiro. Rio + 5 was organized by the Earth Council, a non-governmental organization (NGO) set up by Maurice Strong, who chaired the negotiations at the 1992 Rio Summit. A wide range of partners lent their financial support and their influence to the event, and meetings were held around the world to allow input to the preparations. Four hundred and twenty-two participants from all sectors and geographical regions were invited to share their experiences and to 'contribute to the fifth anniversary review process'. The heads of the United Nations Educational, Scientific and Cultural Organization (UNESCO), the United Nations Development Programme (UNDP), the United Nations Environment Programme (UNEP) and the World Bank all attended, as well as eminent representatives from national and international NGOs, national councils for sustainable development, the private sector and so on.

The forum was criticized by a number of organizations involved in the intergovernmental preparations for UNGASS, who questioned the value of the event and the use of resources it entailed. The Earth Council maintained that Rio + 5 should be seen as complementary, and that the opportunity for 'sustainable development practitioners' to meet before the Special Session would help them to influence preparations more effectively than would otherwise be possible. Whatever the rights and wrongs of this argument, it is clear that Rio + 5 did not have the major impact on UNGASS that its organizers hoped for. Attempts were made during the forum to agree a set of recommendations from delegates, and to get backing for the preparation of an Earth Charter. Both of these proved controversial and in the event no strong endorsement for either could be claimed. The texts made available in New York did not have a strong impact on the intergovernmental process.

Given the number of influential participants, and the unavoidable overlap that many would see between the forum and preparations for UNGASS, the shortcomings of Rio + 5 must have had an impact on the Special Session. For some it was taken as evidence that UNGASS could not succeed, and as a result they did not play as active or constructive a role as they would otherwise have done.

THE COMMISSION ON SUSTAINABLE DEVELOPMENT

The CSD meeting was a more complex process than the intersessional meeting. Dr Mostafa Tolba of Egypt was elected to the chair, and four others joined him in the bureau – Ambassador Bagher Asadi (Iran), John Ashe (Antigua and Barbuda), Monika Linn-Locher (Switzerland) and Czeslaw Wieckowski (Poland).[2] In addition, Dr Tolba invited Ambassador Amorim and Mr Osborn to continue to assist the bureau as friends of the chair.

A number of ministers were present in the first week (mainly environment ministers, and rather more from the North than the South). They gave their political guidance on directions to be explored, and their comments and views were then condensed into summary documents and explored further in the negotiations that followed. The European Union (EU) launched initiatives on fresh water, on energy, and on sustainable consumption and production.

In the second week there was a series of nine half-day dialogue sessions between representatives of the various different major groups and official national representatives regarding aspects of sustainable development of particular concern to them. The intention was to assemble a good record of their achievements and commitments, and to establish a stronger dialogue between governments and all the

During the fifth session of the CSD the European Union launched three initiatives. The CSD showed support for our proposal for activities aimed at an efficient and equitable distribution of water resources and their integrated, sustainable management. Furthermore, concerted action is required to provide for coordinated sustainable energy policies. We will pursue this initiative. Lastly, studies have indicated that 50 years from now the world will need a tenfold increase in eco-efficiency. Economic development without vastly improved efficiency of natural resources and energy use will gradually but inevitably come to a complete standstill. The European Union has proposed to study the feasibility of a fourfold increase in eco-efficiency which should be achieved within two to three decades.

Dutch Prime Minister Wim Kok addressing the General Assembly Plenary on behalf of the European Union

major groups so that each could learn from the others and help to make their different activities more mutually supportive.

Good material was presented at many of the sessions that will be useful for the CSD's future work, but attendance at the dialogues and the quality of the debates were not as high as had been hoped. The dialogue sessions had to be scheduled at the same time as negotiating sessions and other meetings, and they were therefore not very well attended by other groups, particularly government representatives. The extent to which the different major groups were able or willing to interact with each other also appeared limited. The results of the dialogue sessions were distilled into summary reports, with the intention that the issues emerging should feed into and inform the later stages of official negotiation. However, the results were not sufficiently focused or timely enough to have the influence they might have had on the process.

The detailed official negotiations on the text in fact commenced in a number of parallel sessions during the second week of the CSD. The main cross-sectoral issues were negotiated in one group, and the main sectoral issues in another. Forestry issues, organizational issues, and the work programme for future CSD meetings were discussed separately in other parallel groups. Dr Tolba also conducted informal consultations on the text of a possible political declaration to complement the main detailed text and express the political commitment of governments to the main themes of sustainable development.

These complex parallel negotiations were necessary because of the large number of topics to be covered in a limited time. But it imposed great burdens on all the participating countries and organizations, and made it very difficult for any one person to keep abreast of the whole state of the negotiations. Inevitably it made for some fragmentation, with good progress being made on some topics some of the time, but with others becoming stuck in confrontation, or in empty repetition of texts already agreed in Agenda 21 or on other previous occasions.

By the end of the CSD a substantial part of a text for the Special Session to adopt had been agreed, but it became clear that too much work remained for the single week of the Special Session. It was therefore agreed to arrange a further week of informal consultations under the chairmanship of Dr Tolba in the week before the Special Session from 16 to 21 June. During that week many of the outstanding issues were resolved and by the time the Special Session commenced the main open questions concerned major issues on which political leadership and guidance were perceived to be needed to resolve matters.

Some progress had been made by the end of the CSD on the key issues that had been identified at the intersessional. A new global initiative on fresh water was agreed, and some challenging possibilities on energy and transport were identified. Ideas for strengthening studies and dialogue on sustainable consumption and production were launched. On forests and climate change, however, the meeting was not able to do any more than identify and clarify the positions of the different groups, and on poverty and finance for development there was little sign of real movement.

AGENDA FOR DEVELOPMENT

Calls for the Secretary General to prepare an Agenda for Development were made in 1992, after the release of Boutros Boutros-Ghali's Agenda for Peace. A number of countries, particularly from the developing world, wanted to ensure that development issues would not be overshadowed by UN work on peace and security. Governments participating in the General Assembly subsequently decided that they wanted to negotiate the agenda themselves, and established a working group for this purpose in December 1994. On 20 June 1997, after four years of negotiations, the General Assembly adopted the Agenda for

Development as part of the ongoing intergovernmental process of UN reform.

Many developing countries were very disappointed with the final outcome of these negotiations, citing particularly the lack of commitment from industrialized countries to provide financial support and technical assistance to help developing countries in meeting their development objectives. Countries were represented at meetings of the working group by their permanent missions in New York. In many instances the permanent representatives, particularly those of developing countries, also led the delegations to UNGASS. Dissatisfaction over the conclusion to the Agenda for Development process can be seen as a factor in negotiations of key sections of the UNGASS document.

THE SPECIAL SESSION

The Special Session itself commenced on 23 June. At the beginning of the session Ambassador Razali Ismael of Malaysia, the current President of the General Assembly, was elected to the chair. The main session was devoted to speeches by heads of state and government and other high-ranking personages.

On the first morning there were speeches of welcome from Vice-President Al Gore of the United States, and Fernando Henrique Cardoso, the President of Brazil. Other speakers on the first morning included President Mkapa of Tanzania, who then held the chair of the G77 Group of developing countries, President Mugabe of Zimbabwe, Prime Minister Kok of the Netherlands, who held the presidency of the European Union, Dr Song Jian of the People's Republic of China, Prime Minister Blair of the United Kingdom, Chancellor Kohl of Germany, President Chirac of France and a number of other heads of government and state. Other heads of state and government spoke later in the week, and President Clinton of the United States gave an important speech on 26 June.

For the detailed negotiations the session appointed a Committee of the Whole under the chairmanship of Dr Tolba. This committee met in parallel with the plenary throughout the week, and itself set up a number of working groups to continue the detailed negotiations in a similar process to the CSD negotiations. In addition to the official working groups on the different topics three informal Ministerial Working Groups were established to try to achieve some political breakthrough on three key issues that had proved intractable in the official level discussions – forestry and whether or not to embark on

Given the global interdependence recognized at Rio, this meeting not only requires us to reaffirm our previous commitments, but to address a new set of challenges that I pose here as questions:

To governments – 'How will you engage in and fulfil global commitments without fearing that you have forsaken the need to look after your national interests first?' Surely it is not that national interests should be compromised in favour of broad international considerations, but simply that national interests can, and should, be defined in terms that encompass the well-being of other states and peoples as being tied into one's own prospects and prosperity.

To the private sector – 'Are the imperatives of profit, new markets, competitive edge and commercial secrecy so great that you remain reluctant to have an open and responsible dialogue with other stakeholders?'

To members of civil society – 'How do you account for five years of lost opportunity?' You are an essential component in this process as producers, consumers, tax-payers, and as supporters and critics of the governments gathered here today. It is your responsibility to actively participate in sustainable development in your own lives, and to demand no less of your political, economic and social institutions.

Ambassador Razali Ismael, President of the General Assembly addressing the GA Plenary

the negotiation of a convention: climate change and the political steer that could be given to the preparations for the next Conference of the Parties (COP) to the Climate Change Convention; and financial issues and what to do about the decline in levels of aid from North to South.

In addition to the plenary meeting and the official negotiations there were a large number of informal discussions, debates, press briefings and other meetings involving ministers, officials, NGOs and representatives of all the major groups on the many issues related to sustainable development. This continuous networking is one of the most important parts of this kind of conference and played, as always, a valuable service in sharing experience of sustainability problems and solutions throughout the world, as well as exerting indirect influence on the negotiations themselves. This stream of activity was well-reported by the daily NGO bulletin *Outreach*, produced throughout CSD and the Special Session. The Earth Negotiations Bulletin was also published daily, and provided an extremely valuable synopsis of the previous day's events and a useful outline of issues to be addressed in the negotiations. This has proved particularly useful to the authors of this text.

The negotiations during the week of the Special Session itself were disappointing. Agreement was finally reached on all the outstanding points, but on many of the issues a great deal of negotiating time and effort was spent in getting to end-points that did not differ substantially from previous agreements, or represented weak compromises between widely divergent opinions. This lack of a clear common purpose was then made worse by acrimonious debates in several of the groups and by tensions between the representatives of the North, which included a number of ministers, and the South, which was predominantly represented by officials and their permanent representatives in New York. In particular, the Ministerial Working Groups failed to achieve any breakthrough on the three issues they had been set up to address, and had to settle for compromises that were no more ambitious than the positions officials had already identified as fallbacks.

ASSESSMENT OF THE SPECIAL SESSION

On the key issues identified at the start of the intersessional the conclusions on poverty were strong, but were to some extent undermined by the failure to agree positive steps on finance for development. The South had reason to feel let down by the aftermath of Rio where they thought they had been promised more assistance to help them on the path to sustainable development. Instead the five years since Rio have seen steady decline in aid levels from the majority of donor countries. Nothing done at the Special Session gives any prospect of this being reversed in the near future. There was reference to the growing part played by foreign direct investment, but this is largely concentrated in certain areas of a minority of countries, and cannot easily be spread more widely to help poverty relief and sustainable development. Innovative ideas, such as the proposal for an aviation fuel tax or a panel on financial issues, also failed to make headway.

Fresh water is to be one of the main themes of the CSD at its 1998 session, and if the initiative launched at the 1997 CSD session and endorsed by the Special Session is followed through vigorously this could provide an opportunity for bringing about a real global partnership to handle a central issue of sustainable development in a practical way.

On climate change, a little movement was secured at the eleventh hour in the final agreement, building on the speeches from heads of government. Most of the Organization for Economic Cooperation and Development (OECD) countries said they were now ready to consider committing to legally binding targets for significant reductions in greenhouse gas emissions at Kyoto in December 1997. This limited agreement helped to build momentum towards Kyoto, but it still left much to be done before December. On related issues some commitment was secured to promoting more sustainable energy policies throughout the world over the next few years; and new ground was opened for international cooperation on transport.

Although the European push to open negotiations on a convention on forests did not carry the day there was good support for a new Forum to carry forward international consensus, building on the principles and practice of sustainable forestry and to identify elements for a possible future convention. There was also widespread recognition of the looming catastrophe of over-fishing, which is driving many species to the point of extinction, and the need to reinforce and implement more strongly the various international agreements on this subject.

Drawing all this together the session agreed a detailed assessment of progress since Rio and a programme for the further implementation of Agenda 21. It also put forward proposals for streamlining the work of the CSD itself, and a coherent programme of work for that body for the next five years. So, although the progress is not as rapid as some would like and although the political momentum was not strengthened, there is good material to build on in the years ahead.

All of these conclusions are embodied in the Programme for the Further Implementation of Agenda 21, which was the main output of the Special Session. This is analysed in more detail in Chapter 4.

NOTES

1. 38.9 pf Agenda 21.
2. Monika Linn-Locher was replaced as a member of the bureau before the Special Session by Idunn Eidheim (Norway).

3

Great Expectations: Towards Earth Summit III in 2002

Rio + 10 in 2002 will be a bigger day of reckoning than the five-year staging post we have just had. Looking ahead, the world needs to consider how the international situation could be transformed by then and a more effective partnership for sustainable development put in place.

Before 2001, a five-year review of the 1995 Copenhagen World Summit for Social Development is due in 2000, when issues of social injustice, aid and poverty will again be at the top of the agenda. The Secretary General's reform proposals have also canvassed the possibility of a Millennial General Assembly in 2000, which could provide an opportunity for reviewing all the major objectives of the United Nations, including sustainable development.

It is important to establish a coherent international timetable for the next five years as soon as possible so that when we come to Rio +10 there is genuine progress to report. Establishing realistic expectations, a realistic timetable, and a more realistic political commitment to effective follow-up and implementation will be a key part of success.

The programme of subjects now established by Earth Summit II for the annual meetings of the Commission on Sustainable Development over the next five years is an important part of this. It provides an orderly sequence for working through the main topics in a sensible way, with different subjects allocated to different years so that the CSD can focus on the areas in which there are currently gaps, or usefully add to work underway in other international fora.

Programming alone cannot achieve the progress that is needed, however. The essential difficulty with advancing sustainability throughout the world is the tension between short-term and long-term perspectives. In the short term, economic imperatives frequently seem to be of overwhelming importance for individuals and for governments. The longer term perspective that sustainability requires seems like a luxury that cannot be afforded. People feel the need for growth now, for consumption now. The longer-term threats that environmentalists and others perceive may never materialize, or they will be for someone else to worry about when the time comes.

What can the advocates of sustainable development set against this overwhelming pressure? Three things are essential.

- Good information and good science, well disseminated. Everyone needs to know what is happening to the world around them, and continually improve their understanding of trends and processes. Scientific results need to be disseminated widely so that no one can hide behind the excuse of not knowing what is going on and the threats or risks ahead.
- Good economics and policy analysis. Why does development take the paths it does? What motivates people, companies and countries? What measures can alter behaviour in a more sustainable direction?
- Political support for such changes has to be mobilized and made into an effective instrument for countering short-termism? How can all parts of society be engaged in the issues? How can changes in values and lifestyles be brought about? How can countries learn from one another and cooperate together effectively instead of indulging in competitive short-termism and beggar-my-neighbour policies?

The world is getting quite good at the science, the economics and the policy analysis of sustainable development, although there is still more to be done to improve all of these. However, it is not on these matters that Earth Summit II and other international conferences have fallen short. The failure is much more a failure of dissemination of information in a vivid way, leading to a failure to mobilize public opinion and political pressure on a sufficient scale. The key task for those considering future occasions is to improve in this area.

One argument sometimes heard from those opposed to vigorous action in the sustainability field is that the political process cannot buck public opinion, and that if public opinion has moved away from long-term concerns with the environment and sustainability there is nothing that concerned people can do about it until public opinion mysteriously reignites its concern at some point in the future.

As an argument for doing nothing until overwhelming public pressure forces action this is cynical fatalism – true political leadership requires real problems to be addressed even in the absence of strong public pressure. But as a recognition that the United Nations does not exist in a New York vacuum, and that it is not possible to conclude meaningful agreements with real commitments by diplomatic process alone, unless a real popular and political demand in member states exists or has been stimulated it is profoundly true. Meaningful international action on sustainable development can only be built on national and local awareness and action, and on national awareness of the need for international agreement to reinforce and amplify what can be achieved nationally.

From this perspective, the key achievement of Rio was not the texts of Agenda 21 and the other agreements, nor all the surrounding activity in Rio itself. It was the process of building up public awareness of the issues and a popular demand in countries throughout the world for action locally, nationally and internationally. This process took place over several years leading up to Rio. The Brundtland report, *Our Common Future*, gave a persuasive analysis and political lead. A series of regional conferences before Rio helped to define issues and aspirations on the way, and innumerable local and national meetings and discussions built up the pressure further.

Such a process of building up concern cannot of course be manufactured out of thin air. It has to feed on real problems and genuine public concern about them. But the governmental and international organizers of conferences and negotiations can help to ensure that the public concern becomes articulated and organized in a coherent way so that a popular consensus about problems and possible solutions supports and motivates the politicians and officials engaged in the detailed shaping of national policies and negotiation of international agreements. The two-year period of preparation and consciousness-raising for Rio also meant that people and politicians could reach out to each other across national boundaries, learn something of each other's views and opinions, and help to move the positions that their own countries were taking in the interests of securing a consensus.

It was this process of public discussion and debate that was conspicuously lacking on the occasion of Earth Summit II, principally through lack of time, but also to some extent through a kind of confer-ence fatigue amongst some of the principal players, and in the United Nations itself. Earth Summit II had essentially the same agenda as Rio itself, and would similarly have needed a full two years preparation at local, national and regional level to build up sufficient public aware-ness to make significant progress towards new international agreements or action. It would have been difficult in any circumstances to have generated sufficient political or public appetite to attempt such a major build up of activity so soon after Rio. But with significant economic problems in many countries during the mid-90s and with an extensive international programme of other major conferences on other topics between 1992 and 1997 it was impossible for the world to focus attention on Earth Summit II sufficiently far in advance to achieve significant movement among any of the main players. The result was that negotiators at Earth Summit II had little room for manoeuvre, and no political impetus to modify existing positions in the interests of securing new agreement and action.

It could be argued, and was argued, that five years after Rio one should not expect to be making great changes in Agenda 21. The problem was more how to secure its effective implementation – a lower profile task that should not have needed so much political build up, and for which a comparatively short preparation period might have been sufficient.

If Agenda 21 had indeed been advancing steadily throughout the world then a comparatively low profile progress-chasing meeting might have been appropriate. The problem was, however, that in many crucial aspects Agenda 21 was slipping backwards. It was not being implemented properly, and the political deal whereby the North would provide new and additional resources to assist implementation in the South had totally failed to materialize. Bringing Agenda 21 back on course, getting a more effective action programme in place and restor-ing or reshaping the Rio deal between North and South was a much larger task. To have brought that off successfully would have required a bigger build-up to real political movement and commitment. But those who were falling short of Rio aspirations and commitments had little political reason to advertize the fact; and the NGOs and others who ought to have mobilized concern about the failures seemed to have lost some of their energy and bite in many countries.

Looking forward now to Earth Summit III in 2002 it is clear that a much bigger political effort particularly over the two years from 2000 to 2002 will be necessary to secure effective results. Earth Summit II can be regarded as a warning signal that the Rio process and Agenda 21 have been going astray and that much more effort will be needed in the years ahead to put them back on course. Dissemination of the facts about the problems the world faces in an accessible and politically effective way, gathering widespread support for the instruments and measures that will be needed to address these problems and exploring the basis for a new deal between the countries of the world all require time for analysis and political mobilization.

At the level of international agreement the essential task is to explore the basis for any new deal, which can underpin all other agreements. The Rio deal seemed like a good one at the time – new and additional resources to assist the South in adopting more sustainable development policies that would enable these countries to avoid at least some of the development mistakes that the North had made in earlier years, and thus avoid exacerbating world environmental problems. But the Northern countries, with some honourable exceptions, failed to deliver. Countries of the North are widely perceived to preach sustainability, but to lack the political will to make a reality of their environmental and social commitments, both in their own countries and in the assistance they are willing to deliver to the developing countries.

The South, for their part, have made very clear that their priorities are economic growth and poverty eradication, and that they require less preaching from the North and more assistance. In the South's view, environmental initiatives are no compensation for the palpable failure of partnership for development, demonstrated by the growing gulf between the richest and poorest countries, the increasing levels of absolute poverty, and the decline in levels of aid.

For the future, the North needs to forge a credible strategy for promoting sustainable development systematically through all their domestic and international policies and particularly in their relationships with the countries of the South. At the same time the countries of the South, and particularly the dynamic economies that are poised to repeat many of the development and environmental mistakes of the North, need to take on board that sustainability is not a matter of optional environmental frills – it is essential to any sensible path of future development for them just as much as for the richer North.

These are big tasks, which will need long-term strategy, operating at all levels – local, national, regional and international. Governments will need to produce more rounded sustainable development strategies, with the social and economic dimensions given as much weight as the environmental. Local government will need to commit to Local

Agenda 21, which is already very vigorous in many parts of the world. Industry will need to develop further its ability to deliver sustainable growth and protection of the environment.

It will also be essential to make a more determined effort to secure agreement from all the donor countries for a new deal on the financial issues. Perhaps some rapidly developing countries will be moving beyond the need for official aid from developed countries and will be able to rely more on private sector investment and finances. But for others the crisis of poverty and environmental degradation is more acute than ever, and there is still a clear need for a reversal of the decline in aid, more debt relief, replenishment of the Global Environment Facility (GEF), and additional finance for the new water initiative and the objects of the Desertification Convention.

PROCESS

Negotiating in the United Nations is a complex process, in which even expert participants can stumble and in which the less expert can get totally lost and frustrated. Procedures are sometimes laborious and time-wasting, and sometimes jump forward with such speed that important points are lost. Non-governmental participation can be particularly problematic – welcomed and encouraged by some of the participants some of the time, distrusted or excluded by others at other times.

The experience of Earth Summit II shows a number of issues on which improvements might be made:

- analysis, coordination and preparatory work within the United Nations system;
- allocation of tasks and issues to appropriate fora;
- political promotion and animation of the process;
- improved negotiating procedures – how to secure effective and adequate participation of a full range of all concerned government departments, and of all major groups of civil society.

The United Nations system is full of excellent analysis and reports in many different parts of its structure, all bearing on sustainable development in its various aspects. However, coordination and implementation are harder to achieve. The CSD and the Department of Economic and Social Affairs have been able to make some significant improvements in the five years since Rio, and to give sustainable

Finance debates at Earth Summit II were an unedifying dialogue of the deaf. Developed countries' repeated references to the importance of private investment for development inflamed the South's legitimate frustration at declining aid budgets. Moreover, they did nothing to promote constructive debate on ways of meeting the South's diverse environmental and social needs. Aid cannot do everything, and neither can private investment. What is needed between now and Earth Summit III is mature intergovernmental discussion of how the various financial tools available can be best used to do the variety of urgent jobs at hand – different types of environmental protection (from biodiversity to sewage treatment), different priority tasks for development (from health care to export promotion).

Some of this debate would be best conducted under the auspices of individual global environmental conventions, to bring specialist expertise to bear and ensure integration with the wider challenge of implementing them. UNEP and other agencies – including financial institutions such as the World Bank – also have an important part to play. But all this must be overseen by the CSD – the only institution with a mandate to see the connections among the multiple threads of environment and development, and promote overall coordination and consensus. After the limited achievements of Earth Summit II, and the daunting challenges facing us before Earth Summit III, this mandate makes the CSD more important than ever.

Rob Lake, Royal Society for the Protection of Birds

development a key role in the whole structure of UN bodies. The Administrative Committee on Coordination reported to the Special Session that 'it is the collective view of the executive heads of the organizations of the United Nations participating in the Administrative Committee on Coordination that the concept of sustainable development provides an over-arching policy framework for the entire spectrum of United Nations system-wide activities at the global, regional and country levels'.[1] Nevertheless, this still needs to be followed through and implemented in all the relevant parts of the system, and made a key part of the implementation of the Secretary General's reform proposals.

Many of the individual aspects of sustainable development are assigned to particular lead agencies, or to single topic conventions with their own governing conferences of parties. This is frequently the most effective way to handle individual topics, but it makes it difficult for those bodies or those occasions responsible for overview assessments of progress, such as the Earth Summit meetings, to draw all the threads together, and to inject new energy and impetus where it is needed. For an effective overview of this kind it is essential that the process leading up to a summit meeting of heads of government should have the authority, time and resources to obtain progress reports from all the relevant parts of the UN family, and should be capable of taking a broad view of whether or not their individual efforts go far enough towards achieving the overall goal of sustainability or whether they need reinforcing at any point.

There also needs to be a clear view as to which topics should be internationally negotiated in an appropriate forum with binding agreements and commitments emerging, and which topics are best left to national discretion, but on which there may still be a useful place for an exchange of information from one country to another.

This connects with the difficult question of how to get political animation into an overview process such as the Earth Summit. One important requirement is to undertake the assessment of progress as objectively and dispassionately as possible. It had been intended that this should be done during Earth Summit II but in the event there was insufficient time and the meeting started negotiating new actions before the implications of the assessment of progress had been agreed and weighed up. This weakened the effect that a proper assessment pointing up all the failures since Rio might have had in galvanizing countries to new efforts.

The second gap in the Earth Summit II process was the absence of a political animator of the process to play the role Maurice Strong had provided for Rio. The secretariat provided good analytical papers, and the bureau did its best to guide negotiations once they had got under way in New York, but no one had undertaken high-level political consultation and mobilization in capitals in advance or during the process, or built up the interest and concern of the public and the media throughout the world. The world leaders felt no concern or pressure on them to deliver results at New York, and few felt any obligation to do much more than deliver a speech to the Special Session. For the 2002 event a two-year build up coordinated by a strong secretariat could give the whole process much more political authority.

As to the negotiating procedures themselves it would be desirable to seek to identify those who are going to be responsible for chairing and guiding the different parts of the negotiations earlier so that they can prepare themselves more thoroughly, both as to the substance of the issues to be negotiated and as to the positions of the parties. They could then play a more significant role in guiding the debates to a successful conclusion, and ensuring that texts build on previous agreements rather than marking time or occasionally even drifting backwards.

For overview meetings leading to summits it is essential that all the relevant departments and ministers of governments are involved in the preparations. Serious meetings on sustainable development need to involve departments and ministers responsible for finance, development, industry, energy, transport, education and agriculture as well as environment if significant agreements and commitments are to be made. The timetabling of the involvement of NGOs and other major groups could also be improved. It is important that their views are well developed and are given a chance to feed into the official discussions in good time rather than coming on to the scene too late to have any substantial influence.

All the above refers to improvements to the negotiating processes in New York and other international fora. But the key mobilization of opinion and political purpose has to take place at national level. The further development of effective national strategies for sustainable development with the full and effective participation of all the major groups of society is crucial. Awareness of problems and options needs to be built locally, nationally and internationally. All relevant government departments and other major actors need to be engaged. There needs to be sufficient publicity to ensure active negotiation and a determination to achieve positive outcomes. The political message from the top is crucial. Governments need to integrate sustainable development into the heart of their core mission and to drive it forward; at the same time the message from the top needs to respond to the voice of local communities and the people expressing their own vision of a better future.

As we move forward towards 2000 and 2002 three key elements need to fall into place. First, we need a new overview of progress and threats, written in an authoritative but popular way, that will reawaken the citizens and countries of the world to the challenge of the next century. A single author might be commissioned, or a new Brundtland Commission appointed. The report might appropriately be timed to come out in 2000. This would help to capture the attention and imagination of the world in that millennial year, and provide an excellent basis for the build up to another Earth Summit in 2002.

Second, the countries of the North particularly should seek to establish a strategy for the kind of deal involving both public and

This Earth is the only planet in the Solar System with an environment that can sustain life. Our solemn duty as leaders of the world is to treasure that precious heritage and to hand on to our children and grandchildren an environment that will enable them to enjoy the same full life that we took for granted.

Like other nations, Britain is preparing to mark the coming millennium. But the millennium project on which we must all work is to rescue the global environment so that it can nurture life in all our countries for another thousand years. Let us show this week that we have the vision to rise to that task and the commitment to see it through.

UK Prime Minister Tony Blair
addressing the GA Plenary Session

private sectors that they can offer to the rest of the world to secure a more realistic path towards sustainable development in the next century. At the same time the countries of the South, and particularly the dynamic economies that are beginning to face some of the environmental problems associated with rapid development that have affected the North, need to develop a clearer view of what sustainability can do to assist their future development and the implications for them. The International Financial Institutions (IFIs) and such organizations as OECD may be well placed to assist in developing this thinking.

Thirdly, there needs to be clear thinking about the international agenda of the next few years, and which issues could be given additional momentum by the prospect of an Earth Summit III in 2002. These could, for example, include economic issues such as the interaction of trade with the environment; social questions; institutional issues, such as nationalization of the present pattern of international bodies concerned with the environment; and further progress on a range of development and environment issues such as climate change, biodiversity, forests, fresh water and marine resources.

The move from SLUDGE to DREAMS is never going to be easy, but neither is it impossible. It needs steady attention, and long-term planning and above all it needs political leadership and popular support.

NOTES

1.   United Nations General Assembly document A/S-19/6.

Our Mutual Friend: Agenda 21
A Commentary on the Programme for the
Further Implementation of Agenda 21

INTRODUCTION

The main formal product of the Special Session was the text finalized on 27 June under the title 'Programme for the Further Implementation of Agenda 21' (see page 127). This chapter describes how the negotiations went on each part of this document, and provides a brief assessment of the outcome. It takes the form of a commentary, section-by-section, on the Programme.

SECTION A: STATEMENT OF COMMITMENT

The statement of commitment is the political heart of the process. It is intended to embody the core of the political agreement between the participating countries about their future actions.

It was recognized at the outset of the negotiations leading up to the Special Session that political agreement was going to be difficult to achieve on this occasion. There was widespread agreement that the high hopes of Rio embodied in Agenda 21 had only been partially and inadequately fulfilled, but there was much less agreement about what should flow from this. Should countries make new efforts to fulfil Agenda 21 and enter into new and stronger commitments to deliver its aspirations? Or should they recognize that the political appetite for what are seen as difficult commitments on sustainable development has diminished in most parts of the world in the last five years, and that it would be unrealistic to attempt to agree stronger commitments when there has been insufficient political will to deliver the original Rio ones? The story of the negotiations can be read as an initial attempt to agree a stronger commitment that fell back on a reaffirmation of Agenda 21 when it finally became clear that there was insufficient political agreement to sustain the stronger formulation.

Two divergent views about the form of any political commitment had already emerged at the intersessional. Some, principally among the Northern states, argued for a separate and strong statement of political commitment to be prepared that the heads of government could adopt at the Special Session, giving their political endorsement to the conclusions and expressing their commitment to the follow-up. Others, principally among the Group of 77 Non-Aligned Countries (G77), argued that all that was necessary was a short preamble to the main negotiated text.

This issue was unfortunately not resolved, and remained to haunt and disrupt the proceedings right up to the very last day of the Special Session. At the intersessional, the issue was noted but put to one side. It was decided that the task at that stage was to agree the analysis of the situation, and to begin to identify the detailed follow-up action now required. The political elements would be added at the CSD and the Special Session itself. The intersessional, therefore, did no more than identify a few basic elements or building blocks that should appear in the final text of any political statement, chief of which was a recommitment to Agenda 21.

At the end of the intersessional the two co-chairs were invited to consider the matter further and to offer a possible text for a political declaration or preamble before the CSD meeting. After some consideration they concluded that it would be difficult for them to produce a strong draft for a political declaration because there had not been sufficient consensus or political guidance for this at the intersessional. They therefore prepared a shorter statement of commitment in the form of a possible preamble to the main text that they had produced at the end of the intersessional. This was circulated just before the beginning of the CSD.

However, this text did not win much support at the start of the CSD, particularly from the EU and others who wanted a stronger political commitment to be made. As a result, Dr Tolba, with the assistance

of Ambassador Linn-Locher, began a series of informal consultations at
the CSD to identify possible elements for a more extended political
text. At the conclusion of the CSD they circulated a text that was then
subject to detailed negotiation at the Special Session itself. This text
attempted to revive something like the Rio deal. On the one hand it
incorporated stronger and more specific language on the overriding
importance of poverty reduction and of mobilizing new resources to
tackle this, and on the other hand it identified a number of specific
topics, mainly on the environmental side, that might be given priority
for the most urgent action and implementation.

Dr Tolba clearly hoped that this text would galvanize political activity between April and June, so that countries would come to the
Special Session ready to strike significant new deals on these issues at
the political level. It gradually became clear however in the consultations in the week before the Special Session itself that most countries
and their leaders were not prepared to take up this challenge. They
were either unready or unwilling to enter substantive new commitments on the key issues. The majority of countries wanted to do no
more than set out sensible next steps on all the detailed issues in the
course of negotiating the long text and for the political declaration to
confine itself to a general reaffirmation of support for Agenda 21 and
commitment to the programme for its further implementation.

The G77 and others reinforced this view with two procedural
objections to the negotiation, which in the end proved conclusive.
First, they expressed doubts about the informal process that Dr Tolba
adopted at the CSD to produce the first draft of a political declaration.
They objected that the process was not transparent or inclusive
enough. Even when a text was available for negotiation at the Special
Session they preferred to concentrate effort on the negotiations of the
long text. As a result it proved difficult to make headway in the negotiations on the political declaration in the Committee of the Whole, and
the text gradually became longer, with increasing repetition of
elements already agreed in the separate negotiating sessions.

At the same time substantive progress on the main issues in the
various negotiating groups was also proving elusive during the Special
Session. Modest steps forward on some subjects were achieved, but
there were no major political breakthroughs on the most difficult
issues, and above all no progress towards re-establishing the Rio deal.
Some Northern countries made an effort to secure a commitment from
donor countries to reverse the decline in levels of aid, and to set international cooperation for sustainable development on a new and more

positive path again. They tried to convey the vision of inclusivity at the international level. But several major countries (including the USA, Germany, Japan and France) could not or would not move on finance. It became painfully clear to everyone present that the total aid effort from the North is in fact more likely to continue to decline as a percentage of GNP over the next few years, unless much more political will is mobilized.

The United States and some other donor countries argued that the growth of private sector investment flows to developing countries has in recent years become much more significant than official development assistance as an engine of growth and development. While there was general acceptance of the growing significance of this trend, particularly for the dynamic emerging economies, it was also generally recognized that official aid still has a crucial role to play, particularly for the least developed countries in Africa and elsewhere.

Ultimately, therefore, the session felt it right to reaffirm the UN target of raising aid levels to 0.7 per cent of GNP while recognizing the bleak fact that levels of official development assistance are currently declining and that there is little immediate likelihood of this trend being reversed.

All this meant that the mood at the end of the conference was sombre. In a final late night session of the Committee of the Whole it proved impossible to bridge the gap and agree a substantial political declaration on the lines of Dr Tolba's draft. So on the final day the meeting fell back on the brief and modest text embodied in Section A.

Some countries claimed at the end of the Special Session that the meat of the process lay in the more detailed sections on which agreement was satisfactorily achieved, and that nothing of significance was lost by abandoning the longer political declaration. They even claimed that they had never wanted a longer political declaration, which in their view would have distorted the balance of the longer text. Although this may be technically correct, it is difficult to avoid the conclusion that the collapse of negotiations on the political declaration does signify a marked failure of political will to revive the Rio deal and make a reality of it.

SECTION B: ASSESSMENT OF PROGRESS SINCE RIO

Comprehensive reports by the Secretary General on the implementation of Agenda 21 were made available at the start of the intersessional

> The capacity of developing countries to implement Agenda 21 depends critically on increased flows of net ODA [Official Development Assistance] to them. Each decline in ODA, therefore, erodes the capacity of developing countries to implement the Rio agreements and action plan. That is why environmental degradation, which could have easily been prevented, has persisted. Poverty has worsened in some areas, and income inequality within and between countries has widened.
>
> *President of Tanzania Benjamin Mkapa*
> *addressing the GA Plenary Session*

in February 1997. By common consent they were welcomed by delegates as providing an excellent background and basis for forming an agreed common assessment of progress since Rio. After preliminary discussion, the co-chairs produced a first version of a summary assessment at the end of the intersessional. A number of amendments to this were proposed at the CSD, mainly to soften implied criticisms of particular groups of countries. Most countries showed themselves to be more adept at perceiving the motes in others' eyes than the beams in their own. After further discussion the co-chairs provided a further revision that was unanimously agreed at the end of the CSD meeting, and then survived unaltered in the text agreed by the Special Session.

The first three paragraphs of the assessment attempt to summarize changes on the economic, social and environmental fronts. Although recognizing some positive developments, the assessment points up growing problems of poverty, inequality and environmental degradation. Para 12 rightly recognizes the enormous contribution made to sustainable development by all the major groups in society – a contribution forcibly brought home to the official delegates by the very disciplined and positive contribution to the discussions made by representatives of the major groups, particularly during the intersessional and the CSD. The later paragraphs of this section of the text point up the adverse trends on finance, debt and technology transfer.

SECTION C: IMPLEMENTATION IN AREAS REQUIRING URGENT ACTION

The heart of the programme agreed at the Special Session lies in

How can we reconcile our disproportionate contribution to the world's environmental protection and economic growth with our disproportionate impoverishment materially and socially, leading even to the extinction of many of our peoples?

Global trends are dismal, but in the few areas where indigenous peoples are granted respect impressive progress can be made. Indigenous peoples are not peoples of the past but your contemporaries, and maybe your guides towards sustainable futures.

Joji Carino, International Alliance of Indigenous and Tribal Peoples of the Tropical Forests, addressing the GA Plenary Session

Sections C and D. The basic organization and structure of the material in these sections was discussed and broadly agreed at the intersessional in February after delicate negotiations. It was agreed at that time to start in section C1 with major cross-cutting socio-economic themes and the importance of integrating these objectives with environmental concerns in a balanced approach to sustainable development. The text would then go on in section C2 to discuss particular sectors and issues. In C3 financial questions and other means of implementation would be taken up. Institutional questions and the future programme of the CSD would be addressed in the final section, D.

This structure proved to be robust, and the allocation of topics to these different sections survived broadly unchanged through the CSD and the Special Session itself. Para 22 which introduces the whole of Section C, speaks of forging new partnerships and giving priority to the implementation of the whole of Agenda 21 in an integrated way, recognizing the principle of common but differentiated responsibility spelt out in the Rio declaration. The structure proved useful in some ways in organizing the work at CSD and the Special Session, with the sections being assigned to parallel negotiations with different national negotiators in different rooms, and separate members of the bureau guiding the negotiations. However, this led to differences in approach and atmosphere as each section and each topic was treated very much on its own. This meant that it was difficult for participants to keep proper track of the overall shape and balance of the whole negotiation as it proceeded, and try to guide it to a coherent and integrated deal.

It was highly encouraging to hear Secretary General Kofi Annan and others talking at UNGASS in terms of 'governments, business, and civil society'. Further confirmation of this development was the mandate given by the Secretary General and the President of UNGASS to the DPCSD [Department for Policy Coordination and Sustainable Development]. The latter will now work with the WBCSD [World Business Council for Sustainable Development] (and the International Chambers of Commerce), both to formulate ideas and frameworks for public–private partnerships within the UN context, and to identify the framework conditions necessary (and the respective roles and responsibilities of governments and the private sector within these) in order to implement business strategies for sustainable development.

The challenge, of course, is how to translate this into action, so that in another five years' time Earth Summit III will report major progress towards sustainable development. In conclusion, it has to be said that business has in the past found the UN's processes and protocol highly frustrating. The extent to which the UN's structural reforms impact also on this issue will greatly influence the speed at which the new approach outlined above yields implemented frameworks, policies and projects.

Tom Button, World Business Council for Sustainable Development

1: INTEGRATION OF ECONOMIC, SOCIAL AND ENVIRONMENTAL OBJECTIVES

The central tenet of Agenda 21 and the other Rio agreements is that economic development, social development and environmental protection are interdependent and mutually reinforcing components of sustainable development. Text agreed before the Special Session describes sustained economic growth as essential in this context, and stresses the need to share its benefits equitably.

Since UNCED many countries have prepared sustainable development strategies that have proved useful in establishing goals and priorities. At the Special Session it was agreed to call for strategies from all countries by 2002 (when a further five-year review session will take place), as the lynchpin of national efforts to integrate economic, social and environmental priorities. This process should be transparent and participatory, involving all areas of civil society. In addition, it should allow for closer cooperation between governments. While countries agreed general references to quality of life, democracy and equity, a proposal put forward by Switzerland noting the opportunities for job creation while safeguarding workers' rights was opposed by the G77 who were not happy with the allusion to rights in this context. Delegates fell back on a more general reference to Agenda 21, chapter 29.

ENABLING INTERNATIONAL ECONOMIC CLIMATE

During the informal negotiations before UNGASS the United States and the EU pushed for clarification that Principle 7 of the Rio Declaration on 'common but differentiated responsibilities' refers only to environmental issues. The G77 resisted this, but eventually accepted language from Agenda 21. In the final text countries emphasized the importance of international cooperation through dialogue and partnership in order to create an economic environment in which issues of finance, technology transfer and trade can be addressed. This is a necessary element in maintaining global progress towards sustainable development.

ERADICATING POVERTY

Background

Tackling poverty has been widely recognized as a prerequisite for any sustainable development programme. Agenda 21 states:

> It is possible for poor countries to demand that international companies and governments refrain from testing nuclear technology, dumping toxic materials in other people's neighbourhoods, that they do not exploit human and material resources, that they practice fair and just trade, respect the culture and values of host communities and they compensate those they aggrieve. It is possible and reasonable for such companies and governments to reinvest some of their profits in communities where they operate and therefore contribute to sustainable development instead of contributing to poverty and dehumanization of marginalized groups.
>
> *Wangari Maathai, Women's Environment and Development Organization, addressing the GA Plenary Session*

> *A specific anti-poverty strategy is ... one of the basic
> conditions for ensuring sustainable development. An
> effective strategy for tackling the problems of poverty
> should cover demographic issues, enhanced health care
> and education, the rights of women, the role of youth
> and of indigenous people and local communities and a
> democratic participation process in association with
> improved governance. (Agenda 21 3.1)*

The 1995 World Summit for Social Development took the eradication
of poverty as one of its three main focuses, with social integration and
employment. Consideration of poverty eradication in preparations for
the Special Session addressed both the role of international cooperation and domestic action and mobilization of resources.

Much good work has been done in defining poverty and establishing targets for its reduction. The 20/20 initiative, which NGOs pushed
hard to have incorporated in the Social Summit Programme of Action,
calls on both donors and recipients to increase resources allocated to
basic social services. The OECD Development Assistance Committee
(DAC) has developed a set of key objectives that include the halving of
absolute poverty in the world by the year 2015. The United Nations
Development Programme's 1997 Human Development Report deals
specifically with poverty. Its detailed statistical data and reasoned analysis provide a compelling case for concerted global action to eradicate
extreme poverty.

Negotiations

It was widely agreed at the intersessional meeting in February that
poverty was an overriding issue for consideration at the Special
Session. However, given the roles played by other UN bodies and
processes it proved difficult to go beyond endorsing existing commitments. The Bangladesh government submitted a paragraph on access
to micro-credit at the end of the CSD session, drawing on the success
of the Grameen Bank scheme in Bangladesh. Rather surprisingly, given
that the text came from one of their number, the G77 asked that
consideration of this be deferred, as adequate discussion had not been
possible. A final decision on what course of action to take in this area
was deferred to the UN's annual Economic and Social Council
(ECOSOC) session.[1] Priorities that were identified included the
empowerment of people living in poverty; removing barriers that
result in the disproportionate impact of poverty on women; and

> I have heard this week government after government commit itself
> to eradicate poverty. You said the same at Rio, and at Copenhagen.
> Yet, in my country, poverty gets worse, not better. There is no
> chance of credit for young people. It is hard to get a square meal.
> Life is hard, brutish and short: most of my generation will be dead
> before we reach the age of most of you in this room. That is the
> result of poverty, and it is why most young people in countries like
> mine dream of getting away from the God-forsaken place where
> we were born to seek a better future and a longer life in Europe or
> North America. That is the opposite of sustainable development.
>
> *Sheku Syl Kamara, Peace Child Sierra Leone, addressing the*
> *GA Plenary Session*

improving access to resources and facilities necessary to enable people
living in poverty to attain a better standard of living.

Commentary

Consideration of this issue has become bound up with the debate on
aid levels, and acrimony over the perception by Southern countries
that the North has failed to meet its commitments from Rio. As a result
there was little real progress from previous agreements. A UN General
Assembly Special Session to review implementation of the Programme
of Action from the World Summit for Social Development will take
place in 2000.

CHANGING CONSUMPTION AND PRODUCTION PATTERNS

Background

Agenda 21 identified the need for significant changes in resource use
and life-styles, particularly in the North. Finite natural resources and
projected growth in global population mean that severe problems can
only be averted if changes in prevailing patterns of consumption and
production are adopted. Discussion on ways this could be achieved
take two basic forms – increasing efficiency in the use of natural
resources and changing life-styles.

Under the auspices of the CSD a programme of work on consumption and production patterns was led by Norway and Brazil,
culminating in a ministerial meeting in Oslo in 1995 and a workshop
in Brasilia in November 1996. Production and consumption has been
identified as a core issue for UNEP. In addition, the work of UNEP's

Cleaner Production Programme has provided important information on ways in which eco-efficiency can be increased and risks to humans and the environment reduced.

Negotiations

Negotiators spent some time on the perennial question of whether unsustainable patterns of production and consumption are only found in developed countries or are also emerging in developing countries before agreeing a compromise recognising the common but differential problems in this area. A useful list of appropriate steps to strengthen action in this area was agreed.

A proposal for targets that would result in a fourfold increase in energy efficiency over the next few decades, and a tenfold increase by 2050 was one of three EU initiatives presented at the CSD session. This was qualified in negotiations to refer specifically to industrialized countries and to the need for further research to establish the feasibility of these goals and the practical steps to be taken.

Key areas to be addressed were identified, notably the need to shift the burden of taxation to unsustainable production and consumption; the role of business; the need for core indicators to monitor critical trends; the role of education and advertising; and the part governments can play in improving their own environmental performance. G77 countries resisted text proposed by the EU referring to unsustainable lifestyles among higher income groups in developing countries.

In discussion of the CSD's future work programme it was eventually agreed that production and consumption patterns and poverty be overarching issues to be considered each year, with a particular focus on consumption and production in 1999.

Commentary

The text builds on the series of intergovernmental meetings held over the past few years, and provides a strong basis for a systematic examination of ways in which the supply side of the economy could be changed. There is scope for much that the EU put forward to be developed by interested countries through the CSD.

MAKING TRADE AND ENVIRONMENT MUTUALLY SUPPORTIVE

Background

Most governments see the removal of obstacles to trade as being important for spreading the benefits of globalization and economic

A massive education drive must explain to people why consumption patterns have to change drastically, in domestic life and in the workplace, and how it can be done.

The Workplace Eco-Audit, which we have been promoting since 1993, is getting the message through and bringing the workers and employers together to tackle a host of production problems, including the reduction of CO_2 emissions. It also provides a mechanism for monitoring and evaluating progress, and feeding into the national reporting processes called for by the CSD.

Bill Jordan, General Secretary of the International Confederation of Free Trade Unions, addressing the GA Plenary Session

growth around the world. Environmentalists particularly in the North are however concerned that free trade rules are administered by the World trade Organisation (WTO) may undermine laws and agreements to protect the environment.

The WTO's Trade and Environment Committee is intended to address this problem, but has so far made little progress. The division of view has again become apparent over the proposed Multilateral Agreement on Investments (MAI), which is intended to standardize conditions and markets for foreign direct investment. The MAI was initiated through the OECD in 1995. Negotiations on the agreement are currently taking place in a free-standing process, which is due to conclude in mid-1998. Friends of the Earth assert that 'these rules are designed to protect and expand the power of corporations and other large international investors, guaranteeing them a stable investment climate, easy repatriation of profits, open market access, and freedom from any obligation to serve local economic needs wherever they choose to invest'.[2]

Negotiations

The following were among the issues addressed in this section:

CHANGES INTRODUCED BY THE URUGUAY ROUND

Special and differential treatment for developing countries should be fully implemented; in tandem the full integration of developing countries and countries with economies in transition should be pursued. Australia proposed 'the promotion of effective dialogue by the World Trade Organization with the major groups'. This was referred to ECOSOC for a decision.[3]

Final text calls for the relationship between multilateral environmental agreements and the World Trade Organization (WTO) rules to be clarified. A reference to the need for sustainable development and trade liberalization to be mutually supportive (put forward by the United States and EU) was amended and now recommends that trade liberalization should take into account effects on sustainable development.

INTER-AGENCY COOPERATION AND COORDINATION
Strengthened cooperation between WTO, the United Nations Conference on Trade and Development (UNCTAD), the United Nations Industrial Development Organization (UNIDO) and UNEP were called for on issues including trade measures in multilateral environmental agreements, the special needs of small and medium-sized enterprises, and trade and environment issues at regional and subregional levels.

Commentary

In 1992 the impending conclusion to the Uruguay Round of Multilateral Trade Negotiations made it difficult for governments to enter into detailed negotiations on trade issues in preparation for the Rio Summit. Such constraints need not have been felt five years later. However, despite the creation of the World Trade Organization and growing disquiet over the implications of its work and other developments in the intervening period, governments were content to reaffirm the primacy of free trade and its compatibility with actions to protect the environment and social concerns.

Attempts to graft an international legal structure dealing with environmental issues on to existing intergovernmental frameworks set up to regulate trade and investment are proving problematic; and

---

Developing countries, while actively seeking international support and assistance, should fully tap their own potentials. They cannot and should not follow the same old development patterns of developed countries in 'pollution first and treatment later', but rather take the road of sustainable development right from the initial stage of development. Only in this way can the developing countries alleviate poverty and backwardness and instill new vigour into the development of world economy and international cooperation in environmental protection and development.

Dr Song Jian, State Councillor of the People's Republic of China addressing the GA Plenary Session

social concerns inherent in sustainable development provide further complications. The Special Session achieved no more than a stand-off on this issue.

POPULATION

There was no appetite to engage in a major debate on population issues at the CSD and the Special Session text re-emphasises the importance of further promoting the current decline in population growth rates and reaffirmed countries' commitment to the programme of action outlined in the report of the International Conference on Population and Development (ICPD). A Special Session of the UN General Assembly to review implementation of policies agreed at the ICPD will take place in 1999.

In the five years after Rio, globalization is undermining the sustainable development agenda. Commerce and the need to be competitive in the global market, have become the top priority in many countries. The environment, welfare of the poor and global partnership have been downgraded on the agenda. In particular, the 1994 Marrakech Agreements of the WTO appear to be overriding the 1992 Rio Agreements of UNCED and the WTO is now institutionalizing globalization. This globalization process seems to reward the strong and is ruthless in marginalizing the weak. Its paradigm emphasises the gaining of more market share, profits and greed above all else, values that are opposite to sustainable development and global partnership.

Martin Khor, Third World Network,
addressing the GA Plenary Session

HEALTH

Countries reiterated their commitment to implementing the Health for All strategy agreed in Alma-Ata in 1978. Eradication of major infectious diseases, reduction of vaccine-preventable diseases and lowering transmission of other infectious diseases are identified as particular priorities. During negotiations at the CSD the G77 questioned the scientific validity of research showing the harmful effects on human health of exposure to lead and tobacco. These objections were however withdrawn during the week prior to the Special Session.

The text agreed at the Special Session identifies the health effects of lead poisoning, and makes reference to the need for enhanced support and assistance to developing countries in eliminating unsafe uses of lead. The potential risk due to ambient and indoor air pollution, and the adverse impacts of tobacco are also cited as issues requiring action. Health issues should be fully incorporated into sustainable development programmes at every level. Health has not been included specifically in the future programme of work of the CSD, but it will no doubt feature strongly in future work on fresh water, atmospheric issues and other relevant topics.

SUSTAINABLE HUMAN SETTLEMENTS

Approximately half the world's population currently lives in urban settlements; and the total number of urban residents is projected to grow to over 5 billion early in the next century. At the CSD and the Special Session countries called for urgent action to implement the commitments made at the 1996 UN Conference on Human Settlements. The need for decentralization in decision making is stressed.

Local communities and local government throughout the world have made substantial progress since 1992 in developing Local Agenda 21 and other initiatives for promoting sustainable development at local level. Local government was strongly represented at the CSD, with the Special Session urging further support for these arguments. the agreed text suggests that the CSD could establish global targets to promote Local Agenda 21 campaigns and overcome obstacles.

Some countries wanted the Special Session to ensure that the CSD would take a key role in overseeing and coordinating work on these issues. In the end, questions of UN reform being considered elsewhere relating to the future role of the Centre for Human Settlements and the Commission on Human Settlements prevented this from being dealt with fully. The five-year review session after Habitat II in 2001 and the next review of outcomes from Rio in 2002 could be occasions when this can be considered more comprehensively.

> Since 1992, numerous countries have chosen to prepare themselves for the new century by enhancing the status and capacity of local government. More than 70 countries are now engaged in a formal process of decentralization. This is truly a global trend – a trend that reflects the global reality of cooperative governance.
>
> *Councillor Collin Matjila, President South Africa Local Government Association, addressing the GA Plenary Session*

Since Rio, further evidence of the inextricable linkages between health, human well-being and environmental quality have been established. As a result of unchecked environmental degradation, we anticipate an increasing number of human health crises, involving the spread of infectious diseases, growth in malnutrition, and increasing health problems associated with global atmospheric change. Scientists have documented health impacts, such as disruption of endocrine system functions, as a result of toxic chemicals entering into the environment from agriculture and industry.

Proactive environmental health strategies are needed that include adequate research funding to address the linkages between health and environment, with particular emphasis on the common, chronic diseases that affect the poor. We stress that human needs and interests are equally consonant with, and provide compelling justification for, strong environmental protection measures.

Yolanda Kakabadse, the World Conservation Union,
addressing the GA Plenary Session

2: SECTORS AND ISSUES

Throughout the negotiations there was some tension between those, mainly in the North, who wanted to single out some issues for special attention and priority action, and those, mainly in the South, who saw difficulty in prioritising and urged the equal importance of all parts of Agenda 21. The compromise text agreed at the beginning of Section 2 reflects this tension. It affirms the equal importance of the whole of Agenda 21, but then mentions some areas of particular importance (energy, transport, fresh water and marine resources).

FRESH WATER

Background

There is widespread agreement in the development community that the problem of fresh water is likely to be one of the dominant development issues during the next 25 years. Supplies of fresh water are finite, but demands are growing steeply everywhere. Pollution problems are widespread. A number of studies, most notably the joint UN–Stockholm Environment Institute report 'A Comprehensive Freshwater Assessment of the World', have analysed the situation and pointed up the need for new action.

Negotiations

At the intersessional in February it was generally agreed that fresh water problems should be identified as one of the key priorities to be taken up in the Special Session. The EU in particular emphasized this topic, and by the time of the CSD meeting in March they brought this forward as one of its main specific initiatives with a proposal for a new international partnership aimed at building a practical programme of action throughout the world to promote the sustainable use and management of fresh water. The EU proposed that this initiative should be developed further during the next year and that the resulting programme should then be discussed and launched at the CSD 1998 at which fresh water management should be the main sectoral topic for debate.

During the CSD meeting this proposal was first presented informally and discussed in a side meeting. It was then negotiated in detail in the Sectoral Issues Group. The main negotiating problems centred on the desire of the G77 to understand the nature of the initiative and what it could actually deliver to the developing countries. They pointed out that the problems of fresh water are not new, and that many of the measures needed are themselves already widely known and accepted. The problems are the practical ones of establishing the necessary structures on the ground and mobilizing the necessary resources. In particular they urged that unless the developed countries intended to mobilize new and additional financial resources for this programme it could turn out to be no more than a talking-shop.

In the end the developed countries accepted the force of this argument and agreed in the final words of the text of para 35 that the intergovernmental process proposed 'will only be fruitful if there is a proved commitment by the international community for the provision of new and additional resources for the goals of this initiative'. It will be a major task for the preparatory work leading up to CSD 1998 to identify how these resources can be mobilized.

Other issues that were debated include:

INTEGRATED WATER MANAGEMENT

There is widespread professional agreement that integrated water basin management is necessary in order to optimize the use of water in a region. This can be politically difficult to achieve in some parts of the world, and the text was subject to some anxious debate on this issue. Eventually, however, a reasonably firm steer to the virtues of integrated water basin management was agreed in para 34(a).

Let us together decide that in ten years time, every village in the Third World, in Africa in particular, must have its own well or access to drinking water.

Let us together decide to reduce by half, in ten years, the number of urban homes without access to drinking water, or which are not linked to a sanitation network.

Let us decide together to draw up and distribute all over the world, in rural areas and in cities, simple rules for prudent water management.

French President Jacques Chirac
addressing the General Assembly

CUSTOMARY USES OF WATER

In some countries the use of water, particularly from major rivers, is governed by a complex web of traditional and customary rights and practices. Such countries consider it important that the new water initiative should take full account of these customary rights and practices. Other countries believe, however, that such rights may need to be reviewed in the course of establishing more sustainable patterns of water management for the future, and that they should not be given any additional standing or entrenchment by reference to them in an international agreement. The final text recognizes that any new initiative must start from where matters stand at present, but without necessarily stopping there. It refers in para 35 to 'building on existing principles ... and customary uses of water....' This formulation was not acceptable to a number of countries, and on the last day of the Special Session Turkey and several East African countries entered reservations on this point in the final proceedings of the General Assembly.

ECONOMIC VALUATION OF WATER

The EU and many of the developed countries argued strongly that in order to secure the efficient management and use of water it is necessary that it should be regarded as an economic good, and that consumers should be charged an economic price for it so as to discourage wasteful use, and to generate resources for the necessary investments. Other countries were more cautious about the social and political implications of such policies. Nevertheless, the force of the economic argument was generally recognized, and the text (para 34(e)) encourages a progressive move towards economic pricing policies for water as development progresses.

Commentary

Progress was made on fresh water issues at UNGASS. The process was useful in highlighting a major problem that will be at the core of securing sustainable development in many parts of the world in the next decades. The challenge now is to turn this general recognition into an effective action programme, with the commitment and resources of the financial institutions, aid programmes and the private sector all brought to bear. This will be the task of the 1998 CSD. Several intersessional preparatory meetings and conferences have already been arranged, and an Inter-Agency Task Force established.

The CSD's programme for 1998 must aim to improve the efficiency of existing water use; develop water-efficient industry; and ensure that, wherever possible, an economic price is paid for water services. If this can be achieved, future progress could be made using commercial funding as well as official development assistance. This would be a self-sustaining process that could bring about the major improvements that are needed in some regions.

OCEANS AND SEAS

Background

The protection and appropriate management of the oceans and seas has long been recognized as an important part of sustainable development, and one that requires appropriate regional and international agreements. There are already many such agreements covering the law of the sea, protection against pollution, management of fisheries, etc. However, many fisheries in the world are still being over-fished and stocks are declining.

Negotiations

Many NGOs and a number of countries approached the Special Session believing that new initiatives were needed to create new agreements and to establish new coordinating machinery. Others thought that there were sufficient agreements already in place, and that the most important objective was simply to secure effective implementation of existing regional and international agreements.

During the debate on these issues the problems of the oceans and the decline in marine resources were brought into strong focus. These were identified in para 33 of the text as an area where an integrated approach is crucial, bringing together the environmental need to protect the marine ecosystem, the economic need to protect and enhance fishstocks and other marine resources, and the social need to protect the livelihood of fishing communities.

At the end of the debate there was general agreement that this was one of the areas where there are already sufficient regional and international agreements. The task is to implement them firmly and effectively. Para 36 of the text sets this out and gives a particularly strong impetus to the need to eliminate overfishing and wasteful fishing practices, especially in relation to large-scale industrial fishing. It also stresses the need for improved monitoring arrangements.

Commentary

Although there is widespread agreement about the problem of unsustainable management of fisheries it is hard for the international community to get a grip of this in the context of the CSD and the Special Session. Almost inevitably, therefore, the process was driven back on emphasizing the importance of these subjects and commending them to the urgent attention of the various regional and international agreements and the bodies that deal with them. This is worrying given the extent of overfishing in many parts of the world, and the apparent inability of existing agreements to prevent this so far. It will be important for the international community to keep a close eye on progress in these other fora so that if any further action is needed to strengthen existing agreements and machinery this can be considered at CSD 1999 when oceans and seas will be the main sectoral issue.

FORESTS

Background

The loss of forest cover throughout the world, particularly the major virgin forests, was one of the leading concerns at Rio. At that time some wanted a convention to be negotiated to protect forests throughout the world; but others resisted, mainly on the grounds that the management of forests is a national concern and does not require international action. The compromise at Rio was the negotiation of a set of guidelines for the sustainable management of forests – the so-called Forest Principles.

Subsequently, at its fourth meeting, the CSD established an ad hoc Intergovernmental Panel on Forests (IPF), which helped to broaden and strengthen understanding about how sustainable management of forests could best be promoted and encouraged internationally. It finalized its work just before the intersessional meeting in February 1997. Its principal themes were:

- implementation of forest-related decisions of UNCED at the national and international levels;
- international cooperation in financial assistance and technology transfer;
- scientific research, forest assessment and the development of criteria and indicators for sustainable forest management;
- trade and environment in relation to forest products and services;
- international organizations and multilateral institutions and instruments, including appropriate legal mechanisms.

Negotiations

At the CSD meeting and the Special Session the EU, with Germany as a main driving force, pushed strongly for the work of the IPF to be taken forward through the creation of an Intergovernmental Negotiating Committee (INC) leading to a legally binding instrument or convention on all types of forest. Others, notably the G77, China, the United States and Japan, insisted that the need for a convention has still not been established, and that further work on issues not fully addressed by the IPF would be necessary before such a convention could be considered.

The majority of NGOs also opposed the proposal for an INC at this time, fearing that at the present stage of the debate it would enshrine weak standards, favour commercial interests over conservation, and distract attention from the real action needed now to implement existing agreements. They argued instead for new machinery to be set up under CSD to carry forward the work of the IPF and monitor progress.

After lengthy debate the Special Session concluded (para 41) that it would not be appropriate to set up an INC immediately. Instead a new Intergovernmental Forum on Forests will be established to promote and facilitate the IPF's work, to monitor and report on progress, and to consider matters left pending by the IPF. It will also identify possible elements for future international agreements such as a convention, and will report to the 1999 CSD, at which time the possibility of establishing an INC will again be considered.

Commentary

This agreement is a worthwhile compromise. The IPF process is generally regarded as having done good work over the past two years in clarifying issues and in building understanding among practitioners and experts throughout the world about what is involved in securing more sustainable management of forests, and in promoting this. The Special Session's agreement to establish a new forum will ensure that this good work is carried forward and disseminated more widely. Although this falls short of the aspirations of those who wanted to start work on a convention immediately, the new forum will be required to consider the building blocks for a possible future convention so that this can be reviewed more seriously in 1999. By that time it is hoped that there may be more general agreement on the scope and purpose of any convention, and on the resources that might be made available for it. Meanwhile, it will be very important that the forum gives a lot of attention to its monitoring task and to promulgating the results, since forest cover is continuing to decline in most parts of the world, and it is only by giving widespread publicity to this that effective political awareness and concern can be brought to bear on the decisions needed in 1999.

ENERGY

Background

Achieving the best utilization and management of energy is a crucial aspect of sustainable development. Economic growth and prosperity have long been associated with use of energy. But the growth in the use of energy has itself brought problems of pollution and resource depletion that could become critical for the world if past patterns of development are repeated in the developing world.

More recent changes in the developed world encourage the hope that a transition may be possible to less energy-intensive patterns of development in the future. If such transformations can be continued in the developed world and promoted in the developing world a more sustainable future energy path might be achievable.

Negotiations

The EU identified energy as one of the key issues for attention at the intersessional in February. They saw this as a key area in which the environment and development agendas needed to be brought together to achieve more sustainable development. At the CSD in

March they proposed an initiative to establish a new process for the promotion of a sustainable energy future under the aegis of the CSD. Key aspects of this should in their view include full economic pricing of energy, including phasing out of subsidies and internalization of environmental costs, and promotion of renewable sources of energy and energy efficiency. Other Northern countries gave strong support to this approach. NGOs also pressed for sustainable energy policies, but stressed that this should include the phasing out of nuclear power.

The G77 and China gave much greater emphasis to the social and economic objectives of providing adequate modern energy services, especially electricity, to all sections of their populations, particularly in rural areas. Subject to this the majority of the G77 were not opposed to the Northern proposals. There were different strands of opinion within the G77 on this issue, however, and the OPEC countries in particular were suspicious of the European initiative. They took the view that energy issues were adequately dealt with in other fora, and opposed the creation of a new process under CSD.

After lengthy negotiations up to the end of the Special Session a text was agreed (paras 42–6), including in modified form most of the elements that both the North and the G77 had wanted, except that the international process proposed by the EU will take the more limited form of a two-year preparatory process for CSD 2001, when energy is to be a main topic.

Commentary

The agreed text is a good one that has something for everybody. The need for extending energy services more widely in the South is fully recognized, but so too is the need for providing energy throughout the world in a more sustainable way. There is good support for economic pricing policies, and for renewable and low pollution forms of energy. The new CSD-based process proposed by the Europeans was whittled down, but even in its more limited form as an agreement to have a two-year preparatory period on energy issues leading up to the CSD in 2001, there should still be time for in-depth studies and discussions to take place. There is useful support for strengthening coordination on energy issues within the United Nations. It is important that the studies and meetings needed be planned well in advance so that the two years of energy-related work leading up to CSD 2001 will be more likely to achieve some concrete results.

TRANSPORT

Background

Transport is generally regarded as one of the most critical areas for sustainable development. It is one of the fastest growing sectors of the economy in all parts of the world, and is strongly linked to economic and social aspirations, but it takes a growing share of energy consumption, and is a major source of CO_2 and other polluting emissions.

Negotiations

The EU brought forward a number of specific proposals at CSD, including the promotion of integrated transport planning, phasing out lead in gasoline, measures to eliminate subsidies and internalize environmental costs, and a specific proposal for an aviation fuel tax. The G77, and in particular the OPEC countries, were not enthusiastic about strong language on economic instruments and argued for weaker formulations on most of these points. Debate continued longest on the proposed aviation fuel tax, but when it became clear that this was opposed by the United States and Japan, as well as the G77, it was not possible to sustain a reference to this, and the final text merely refers to ongoing work on the use of economic instruments in the International Aviation Organization.

Commentary

Although transport is one of the key issues for sustainability it has commonly been regarded as primarily a national and regional concern. The international community has not yet found an effective way to tackle transport at the global level, although there are some issues, such as the aviation fuel tax proposal, that could only effectively be handled internationally. The final text on this occasion is somewhat disappointing, particularly by comparison with the more progressive text agreed at Habitat II in 1996. The EU and interested NGOs had not done enough work to gather support for their key proposal for an international aviation fuel tax. There will need to be much more thorough preparation for the CSD discussion of this in 2001, perhaps in the context of the two-year preparations for the energy topic.

ATMOSPHERE

Background

There are three main groups of concerns about the atmosphere, each of which is handled at the international level in a different way. The largest problem is climate change, handled by the Conference of the Parties of the Climate Change Convention, which met at Kyoto in December 1997 to review progress and agree new targets and measures. Ozone-depleting substances are dealt with under the Montreal Protocol. SO_2 and NO_X pollution is dealt with under regional agreements on transboundary pollution.

Negotiations

Brief references to ongoing work in other fora on ozone depletion and transboundary air pollution were quickly agreed. The main attention in the negotiations centred on the forthcoming climate change negotiations in Kyoto. Some countries argued that the Special Session should do no more than urge a positive outcome to the Kyoto meeting. Others wanted the Special Session to give a stronger signal to the Kyoto meeting about the outcome desired from that meeting.

At the CSD four possible recommendations to the Kyoto meeting were identified. The EU urged a commitment to seeking targets for developed countries to reduce greenhouse gas emissions by 15 per cent from 1990 levels by 2010. The Small Island States wanted the developed countries to go further and reduce emissions by 20 per cent by 2005. The United States and Japan merely recommended appropriate targets, and wanted to include some reference to action by developing countries, including joint implementation. The OPEC countries and others wanted no recommendation at all about targets. The NGOs urged the same target as the Small Island States. Throughout the CSD all parties continued to affirm their opening

With 4 per cent of the world's population, we produce 20 per cent of its greenhouse gases. Frankly, our record since Rio is not sufficient. We have been blessed with high growth for several years, but that has led to an increase in greenhouse gas emissions in spite of the adoption of new conservation practices. So we must do better, and we will.

US President Bill Clinton addressing the General Assembly

positions and declined to negotiate before heads of government and ministers had spoken at the Special Session.

During the Special Session several heads of government spoke about the crucial importance of this issue and indicated the steps their countries were taking to deal with it and their aspirations for agreement at Kyoto. For the United States President Clinton's speech indicated some movement towards a more ambitious target.

An informal ministerial group was constituted during the Special Session that endeavoured to find some common ground for the text on Kyoto objectives, but was unable to do so. Finally, however, a text was agreed on the last day (paras 48–54) that expressed the partial agreement so far achieved, and indicated the directions in which the Kyoto negotiations should look for solutions.

> There is something perversely unethical about the 'polluter pays' principle, for its reverse logic is that he who can pay can also pollute. This is exactly what happens in the way countries in the North are buying 'forest sinks' in the South (as Norway is doing in Burkina Faso) so that the North can go on polluting.
>
> *Yash Tandon, International South Group Network*

Commentary

The Special Session achieved a useful preliminary airing for the issues on climate change that were to be addressed at Kyoto, and helped to raise consciousness throughout the world about the crucial importance of that negotiation and the positions that the different countries have taken. It was never realistic to expect the Special Session to go much beyond that or to reach any preliminary agreement. However, in para 52 the text did embody some useful pointers towards some key issues that would need to be covered and the direction of the movement that would be needed by some countries by December 1997 if agreement was to be reached. It helped to build momentum towards the eventual Kyoto agreement.

Toxic Chemicals and Hazardous Wastes

Toxic chemicals and hazardous wastes are dealt with in various international fora, and in particular by UNEP and the Basel Convention. Chemicals in particular had been extensively reviewed at the Governing Council of UNEP meeting just before the CSD meeting, where timetables for negotiation of conventions on prior informed consent and persistent organic pollutants had been agreed. At the CSD it was therefore generally accepted that there was not much more for the Special Session to do on these subjects at this stage. The issues were accordingly handled mainly in an informal contact group that produced texts (paras 57 and 58) supportive of and reinforcing the work going on in those other fora. These texts were accepted without question or amendment by the CSD and the Special Session.

RADIOACTIVE WASTES

Background

Most countries had not expected the management of radioactive wastes to be a contentious issue at the CSD and the Special Session. There are well established principles and practices in this area already, and a comprehensive convention on the Safety of Spent Fuel Management and the Safety of Radioactive Waste Management is currently being negotiated under the auspices of the International Atomic Energy Agency (IAEA), so that many thought it would be redundant for the CSD and the Special Session to deal with these subjects as well.

Negotiations

In the event a few countries wanted to go more deeply into the subject. This was partly prompted by a current dispute concerning a proposal by Taiwan to export low-level radioactive waste to North Korea for disposal. This was strongly objected to by South Korea who sought to have the international community reinforce the proximity principle that waste should be disposed of as near as possible to its point of origin to bar such export for disposal.

After lengthy discussion at the CSD and the Special Session a text was agreed (para 60) that reinforces the convention being negotiated under the IAEA, urging its early adoption and giving a qualified endorsement of the proximity principle. In final discussions additional text (para 61) was introduced at the request of the G77 to support the clean-up of sites contaminated by past nuclear activity, and calling for health studies around such areas.

Commentary

Radioactive waste is always a potentially explosive topic at international meetings because of the deep feelings and concerns it arouses, and because of the deeply entrenched positions of all the parties. In the event, although it took up a lot of time at CSD and the Special Session, it was well handled and a useful text emerged. The strong endorsement of the new convention currently being negotiated under the IAEA should help to bring it to an early conclusion and encourage early ratification. The emphasis on clean-up of contaminated sites and appropriate health studies should help move action forward on this front.

LAND AND SUSTAINABLE AGRICULTURE

The promotion of more sustainable patterns of agriculture is one of the key elements of sustainable development. What this requires had, however, already been comprehensively reviewed by the international community at the World Food Summit in Rome in November 1996. There was agreement at the CSD and the Special Session that in general it should be sufficient to reaffirm the conclusions of that conference. Paras 62 and 63 of the text do this.

Two significant points of emphasis were included. First it was agreed at EU insistence that at CSD 1998 the issues of sustainable agriculture and land use should be considered in relation to fresh water since these land use and management issues are a crucial part of securing sustainable management of water resources. The G77 only agreed to this with some reluctance because they are apprehensive of an international process on water getting into delicate and contentious national issues on land tenure, use and management.

Second, on G77 insistence, it was agreed to include a reference to the need to consider further the implications of the Uruguay Round in relation to agriculture and its effects on developing countries. The EU only agreed to this with some reluctance because of its implied criticism of the effects of the Common Agricultural Policy.

DESERTIFICATION AND DROUGHT

Background

Problems of desertification, soil degradation and drought are becoming increasingly severe in many parts of the world. This concern is felt particularly strongly in Africa, and lay behind the urgent plea that the G77 made at Rio for a convention to be negotiated to combat desertification. That convention was accordingly negotiated and came into force in December 1996. The first Conference of the Parties was held in September 1997.

Negotiations

The G77 wished to use the Special Session to secure a strong commitment from the North to help implement the convention and in particular to commit new and additional resources to support this either through the new global mechanism envisaged by the convention or by other means. The North agreed that the issue was important, and that the new global mechanism should promote the mobilization of resources to support the objectives of the convention but were unwilling or unable to commit to additional resources during the Special Session. The G77 were deeply disappointed by this and expressed their disillusion with the outcome in the final session of the General Assembly.

Commentary

The G77 had signalled at the outset of the negotiations in February that they attached great importance to effective implementation of the Desertification Convention and to the provision of resources for it. It was a pity that the North proved unable to bring forward any positive proposals in time to respond to this plea.

> Producing food entails the sustainable care and use of natural resources especially land, water and seed. Farmers who work the land must have the right to practice sustainable management of natural resources and to preserve biological diversity. Genetic resources are the result of millennia of evolution and belong to all of humanity. Farming communities must have the right to freely use and protect the diverse genetic resources, including seeds, which have been developed by them throughout history.
>
> Denise O'Brien, World Sustainable Agriculture Association, addressing the GA Plenary Session

BIODIVERSITY

Loss of biodiversity in the world is a serious threat to sustainability. At international level negotiations about this issue are mainly dealt with by the Conference of the Parties to the Convention on Biological Diversity. At the CSD and the Special Session it was generally agreed that it would be sufficient to endorse and reinforce the main decisions taken by the successive meetings of the Conference of the Parties and in particular the Jakarta Mandate. The agreed text (para 66), which is based on a G77 proposal, does this.

SUSTAINABLE TOURISM

Tourism is one of the fastest growing sectors of the world economy, and one that can have serious impacts on the environment and the sustainability of local economies and ways of life. The CSD and the Special Session agreed that it will be necessary to give increasing attention to this subject in the future. The CSD was accordingly requested to develop an action-oriented programme on sustainable tourism. It was also agreed that this should be a main topic for discussion by the CSD in 1999.

SMALL ISLAND DEVELOPING STATES

The CSD and the Special Session reaffirmed the commitment to the Programme of Action for the Small Island Developing States. The CSD also agreed the modalities for a full review of the programme in 1999 including a two-day special session of the General Assembly.

NATURAL DISASTERS, AND OTHER DISASTERS

The CSD and the Special Session reaffirmed the importance of planning properly to reduce the impact of natural disasters, and of major technological and other disasters with an adverse impact on the environment, and of promoting international cooperation on this.

3: MEANS OF IMPLEMENTATION

FINANCIAL RESOURCES AND MECHANISMS

Background

An acknowledgement of the need for more financial assistance from North to South, coupled with cooperation in other key areas was understood by many observers and participants to be the lynchpin of what has been termed the 'Rio deal'. Throughout preparations for the Special Session G77 representatives referred repeatedly to the failure of Northern countries to move towards the agreed Official Development Assistance (ODA) target of 0.7 per cent of GNP as evidence of their lack of commitment to the Rio agreements. They also successfully insisted on including references to the need for international cooperation and assistance in many of the specific sections of the text. The United States and EU, reacting like the cook when Oliver Twist asked for more, made frequent reference to the importance of private investment flows to developing countries and to the overriding significance of domestic resources.

Negotiations

The following issues were addressed:

ODA AND PRIVATE FINANCE

Wrangles over the relative importance of official development assistance, private direct investment and domestic resources took up much of the time in negotiation of this section. Dutch Premier Wim Kok, speaking for the EU in the General Assembly, stated that 'Foreign direct investment to developing countries has multiplied sixfold during the nineties, but it still reaches too few recipient countries.' An eleventh hour attempt by Dutch Development Minister Jan Pronk to move beyond this impasse failed when both Germany and France, as well as Japan and the United States, spoke against an EU proposal that the decline in ODA should be reversed from the year 2000. The final text falls back on previously agreed language. 'Adequate' replenishment of the Global Environment Facility and the International Development Association in forthcoming rounds are called for.

UN SYSTEM CO-ORDINATION

The joint World Bank–IMF Heavily Indebted Poor Countries (HIPC) Debt Initiative is commended in the agreed text; additional financial resources for its implementation are called for. The World Bank, IMF and UNCTAD are invited to consider the links between indebtedness and sustainable development in developing countries.

INTERGOVERNMENTAL PROCESS ON FINANCE

Proposals for the creation of an Intergovernmental Panel on Finance were strongly pushed by a number of NGOs, and supported by the United States and Norway, among others. The rationale for this was that this panel could consider the policy implications of the work of the existing CSD Expert Group on Financial Issues in Agenda 21, and could report to the CSD. Its remit would include consideration of Foreign Direct Investment, the quality and quantity of aid, and new financial mechanisms for sustainable development and ways in which these could be implemented. This initiative was opposed by the G77, and in large part also by the EU. A decision on taking this forward was eventually deferred to the 1997 ECOSOC session for consideration.[4]

Meanwhile it was agreed to recommend that UNCTAD, the World Bank and IMF should consider sharing studies on such mechanisms with the CSD.

INTERNATIONAL AVIATION FUEL TAX

NGOs, with some support from the EU, called for consideration of a global tax on aviation fuel, revenue from which would be hypothe-cated to ODA for sustainable development. A reference remained in the text until the penultimate day of negotiations, when Argentina opposed its inclusion. An allusion to the concept remains in a call for studies on 'the use of economic instruments for the mitigation of the negative environmental impact of aviation in the context of sustain-able development'. The EU insisted that the tax be mentioned in a footnote to the document.

Commentary

There were few new ideas, and very little room for manoeuvre in the negotiation of the finance section of the text. OECD countries did not come to the Special Session with any new line that could break through the resultant stalemate. Although the Dutch government hosted a meeting of EU development ministers to consider ODA issues in advance of Earth Summit II they were not able to develop a new consensus.

Repeated emphasis by donor countries on the growing signifi-cance of private investment flows did not placate the G77 and China. African states, notably Tanzania, who held the G77 presidency, made it clear that the decline in aid since Rio should be seen as a broken promise, and a measure of the lack of commitment by Northern countries to sustainable development.

We need to mobilize political will to understand the real issues, and the real issues for the South are implementation issues. There is a need for political will that starts with the grass roots people of the North and works up to the politicians and political agencies to the international political level: the G8, the World Bank, the IMF and the Paris Club. You know when the North deals with the South on financial issues they always have their 'clubs'. But we in the South always wind up negotiating as individual countries. A Panel on Finance will not provide political will, only more bureaucracy.

Ambassador Mytusa Waldi Mangachi of Tanzania

The EU may explore the feasibility of introducing an aviation fuel tax within its borders. Whether resultant funds could be earmarked for bilateral development assistance is unclear. The United States has indicated that $100,000 could be made available to establish a forum to address the areas outlined above.

TRANSFER OF ENVIRONMENTALLY SOUND TECHNOLOGIES

Background

Access to information and environmentally sound technologies were confirmed at the Rio Summit as essential requirements for sustainable development. Transfer of knowledge and technology to developing countries was closely linked with increased financial assistance by most commentators as part of the 'Rio deal', and would be necessary for such countries to raise environmental standards. However, it has not proved possible over the past five years for the Commission on Sustainable Development to make progress in realizing these objectives. Debate at annual CSD sessions and at intersessional meetings focusing specifically on technology transfer has become increasingly acrimonious. Northern governments have insisted that the most advanced information is developed and held by the private sector and so it is not in their power to make it available on terms that would be acceptable. The South views this as a retreat from a prior commitment and has pushed for action by the North to find ways to increase their access to all appropriate technologies.

Negotiations

Text agreed at UNGASS recognizes that much of the most advanced environmentally sound technology is in the hands of the private sector. However, governments can play a catalytic role through the use of new

> It has become fashionable to say that governments can do very little, and that all power now lies with unaccountable multinational companies and institutions in a newly globalized market. But let that not disguise the power and accountability which you, together, hold to impose environmental and social limits, controls and standards.
>
> *Dr Thilo Bode, Executive Director of Greenpeace International, addressing the GA Plenary Session*

tools such as 'green credit lines' and the transfer of privately-owned technology on concessional terms. Further studies of ways in which governments and public institutions could generate publicly-owned technologies are called for. In addition, centres for the transfer of technology should be created. Relevant parts of the UN system should cooperate, including regional commissions. South–South cooperation should be strengthened, and trilateral arrangements involving donor countries and international organizations should be supported.

Global information and telecommunications networks are identified as a significant tool for technology transfer that could be further explored and enhanced.

Commentary

Debate of this section retrod ground already covered at annual CSD sessions and during the difficult negotiations over the UN Agenda for Development, concluded a few days before the start of the Special Session. The text does not contain any significant addition to previous agreements.

CAPACITY-BUILDING

The importance of building the capacity of developing countries to design and implement sustainable development strategies is acknowledged in Agenda 21. The United Nations Development Programme has overseen the successful Capacity 21 initiative over the past four years. This was originally given a fixed life span, but UNDP has sought funding to continue its operations until 1999. Text from the Special Session reaffirms the importance of capacity-building, and emphasizes the link between enhanced capacity and the ability of developing countries to absorb and utilize new technologies. As with technology transfers, more South–South cooperation is called for including arrangements involving developed countries and relevant international institutions.

SCIENCE

Text agreed calls for significant increases in public and private investment in science, education and training, and research and development. Greater cooperation and improved access to scientific information related to the environment and sustainable development are necessary.

EDUCATION AND AWARENESS

In Agenda 21 references to education as a support for sustainable development are numerous and spread throughout the text: at Earth Summit II they were relatively few, and largely confined to one mild if supportive section. An effective educational system is described as a fundamental prerequisite for sustainable development in the agreed text. Emphasis is placed on the empowerment of vulnerable and marginalized groups that education can offer, and on the importance of lifelong learning as well as the formal education structure.

INTERNATIONAL LEGAL INSTRUMENTS AND THE RIO DECLARATION ON ENVIRONMENT AND DEVELOPMENT

The 27 Principles of the Rio Declaration provide a useful reference point for national and international laws on sustainable development issues. At the national level, the Rio Principles have been cited increasingly as their effect in the international arena has become more apparent. Agenda 21 and the Rio Declaration are seen as agreements and formulations governments have signed up to, notwithstanding the fact that they are not directly enforceable.

Text agreed at the Special Session calls for regular assessment of their implementation, and highlights Principle 10 in particular, which deals with access to information and public participation in decision making. Development and codification of international law related to sustainable development should continue, as appropriate. A proposal from the Norwegian government for a reference to the need for international law regarding liability and compensation was not accepted.

INFORMATION AND TOOLS TO MEASURE PROGRESS

Background

Collection of authoritative data and the use of relevant indicators are necessary in order to assess whether progress is being made towards the objectives outlined in Agenda 21. Work on development of a set of core indicators has been carried out under the auspices of the CSD for the past few years. Although the principle is widely supported, problems arise given the difficulty in identifying generic measures that are universally applicable and still have relevance in individual countries.

The good examples set by NGOs and enthusiasts in the formal education sector continue to be scattered, and dependent on dedicated individuals whose influence is limited in both time and space. If we have to wait until these contributions have reached a critical mass sufficient to tip the system we may have to wait a long time. New policies need to penetrate into the education establishment (both in formal and non-formal sectors) if they are to reach the bulk of the public through mainstream education and public example: at CSD meetings the education establishment is usually conspicuous by its absence.

Examples of good and successful practice are needed, which illustrate the economic benefits of an environmentally competent citizenry as well as the environmental ones, but CSD and others must create a socio-political climate in which good practices will spread, in which professional educators will see education for sustainability as contributing to professional status and esteem, and in which decision makers see the right educational strategy as an integral part of attaining sustainability objectives.

John Smyth, World Conservation Union (IUCN)

Negotiations

Text in this section was agreed at the CSD session. Areas highlighted included the use of Environmental Impact Assessments; the set of indicators that the CSD will release by the year 2000; and the importance of national reports on Agenda 21 implementation.

Commentary

Many states are wary of measures that suggest comparisons between the records of countries – such concerns are reflected in the final text. However, considerable progress has been achieved in this area, and it has been suggested that the package of indicators the CSD will make widely available in 2000 could assist in the preparation of national reports to the Commission in future.

SECTION D: INTERNATIONAL INSTITUTIONAL ARRANGEMENTS

1: GREATER COHERENCE IN VARIOUS INTERGOVERNMENTAL ORGANIZATIONS AND PROCESSES

Background

The CSD's work over the four years since 1993 has been considerably strengthened by the system of task managers, which was designed to involve relevant bodies of the UN system according to their expertise. This was overseen by the Inter-agency Committee on Sustainable Development (IACSD), a subsidiary body of the UN Administrative Committee on Coordination. However, the record of the task manager system has been uneven, with some UN agencies playing more constructive roles than others. Of concern to many NGOs has been the perception that the International Financial Institutions and the WTO have not been fully engaged in the CSD's activities.

Problems have also arisen from the secretariat arrangements made for the convention processes resulting from UNCED and those of relevance set up before the Rio Summit. These are located in countries around the world, which causes great problems in achieving adequate coordination. UNEP has a responsibility for promoting such linkages, endorsed at its 19th Governing Council in January 1997.

Consideration of the need for system-wide coherence in the UN was made more difficult by the impending release of the Secretary General's reform proposals (17 July 1997). More cynical observers concluded that this constituted a useful excuse for avoiding a far-reaching discussion in this area.

Negotiations

The following were given particular attention:

CONVENTION SECRETARIATS

The desirability of co-location for the convention secretariats is recognized. Improved scheduling of meetings, attention to national reporting requirements and the participation of governments at an appropriate level are also called for.

INTER-AGENCY COMMITTEE ON SUSTAINABLE DEVELOPMENT

The task manager system should be strengthened.

GREATER REGIONAL FOCUS IN THE WORK OF THE CSD

UN regional commissions are preparing reports on their priorities for the future as part of the ongoing UN reform process. Without prejudicing this, the Special Session recommended that the CSD promote increased regional implementation of Agenda 21 in cooperation with the regional commissions and other relevant regional organizations. (The modalities of this are addressed more fully in section D4 (page 66) on methods of work of the CSD.)

Commentary

It has proved very difficult to achieve an effective compromise between mechanisms designed to facilitate progress in specific areas (biodiversity and climate change, for example) and the need for integration and coordination of all these disparate elements in a cohesive structure. The CSD task managers system has had its successes, and it recognizes the distinct roles played by the various bodies while drawing them into more collaborative work under the auspices of the Commission. It will be interesting to see how this is developed given the CSD's more focused programme of work over the coming four years, and also whether this logic can be applied to the convention processes.

2: Role of Relevant Organizations and Institutions of the United Nations System

Background

The 19th Governing Council of the United Nations Environment Programme (27 January to 7 February 1997) brought underlying tensions on the governance of the organization to a head. The influence of the Committee of Permanent Representatives over day-to-day management of UNEP was called into question by environment ministers attending the Governing Council – this division was characterized by many as a North–South split, although in a number of instances there was disagreement between individual countries' foreign offices and environment departments. A Nairobi Declaration was agreed, which spelt out UNEP's future role.[5] However, funds were withheld by a number of Northern countries pending resolution of the dispute over management of the organization. An emergency Governing Council session was held on 4 April at which these issues were resolved to the satisfaction of the environment ministers present.

Negotiations

Difficulties apparent during and after the UNEP Governing Council were not far from the surface in negotiation of this section at the CSD meeting. The G77 did not want reference to the Governing Council decision of 4 April 1997 as they felt it did not add to the decisions relating to governance taken in the Nairobi Declaration. The EU and other countries that had pushed for the subsequent decision insisted that it should be referred to and this was finally accepted.

It has been suggested that Agenda 21 does not define UNEP's role clearly enough, and that as a result it has been marginalized in the follow-up to the Rio Summit. Text in this section is intended to clarify the central role UNEP should play, in line with the thinking behind the Nairobi Declaration. UNEP's responsibilities as the 'leading global environmental authority that sets the global environmental agenda' are spelt out. These include strengthening international environmental law and developing coherent links between environmental conventions.

Reference is also made to the coordinating roles of other UN agencies and programmes: UNDP for capacity-building; UNCTAD for trade and development; the WTO Committee on Trade and Environment, UNEP and UNCTAD for trade and environment (with a coordinating role to be played by the CSD); the World Bank for strengthening the commitment of the international financial institutions to sustainable development; and the importance of putting into operation the global mechanism of the UN Convention to Combat Desertification.

Commentary

UNEP has had a difficult time since Rio, and although its renewed mandate and the new governance decisions of its Governing Council

> Global environment protection and sustainable development need a clearly audible voice at the United Nations. Therefore, in the short term, I think it is important that cooperation among the various environmental organizations be significantly improved. In the medium term this should lead to the creation of a global umbrella organization for environmental issues, with the United Nations Environment Programme as a major pillar.
>
> *Chancellor Helmut Kohl addressing the GA Plenary Session*

and the Special Session should help it to recover its position the outcome remains unclear.

The upshot of the various meetings in 1997 has largely been to preserve the status quo over organizational matters related to sustainable development. The central coordinating role of the CSD has been reaffirmed, and its work programme streamlined for the next five years. Its task manager approach is endorsed, but it remains to be seen whether it can create real cooperation with other agencies on key cross-cutting themes, and whether countries can achieve integrated involvement of all relevant ministries and major groups as well as environment departments and NGOs in the work programme.

3: FUTURE ROLE AND PROGRAMME OF WORK OF THE COMMISSION ON SUSTAINABLE DEVELOPMENT (INCLUDING THE MULTI-YEAR PROGRAMME OF WORK FOR THE PERIOD 1998–2002)

Background

At its organizational session in 1993, the then Chairman of the Commission, Ambassador Razali of Malaysia, placed strong emphasis on the integration of environment and development as key to the work of the CSD. It was to be the focus in the UN system for political and practical consideration of how to realize the Rio commitments.

During three sessions (1994–6), a review of implementation of the whole of Agenda 21 was attempted. Nitin Desai, UN Under Secretary General for Sustainable Development, established a small but highly effective secretariat for the commission, and the task manager system (page 60 above) cemented links with other relevant UN departments and bodies. A strong input by those outside central government in these annual meetings was actively sought by the secretariat, and an increasing range of government departments were represented in national delegations to reflect the issues to be negotiated.

In its substantive work the CSD achieved some notable successes, including high profile work on forests and patterns of consumption and production. It has also come to be seen as an example of ways in which other parts of the UN could play a catalytic role and engage a wider range of participants through innovative approaches and ways of working. However, many of those involved felt frustrated by the weighty agenda to be covered at annual sessions, and by perceived duplication

In many ways, the interagency coordination system, working with the task manager system, has not created a threat to those agencies [such as UNEP] at all. By not sending someone from the outside to investigate their agency, but instead inviting them to report to the CSD themselves, we have not questioned or challenged agency policies nor the mandates of agencies. In fact, we have reinforced them by giving those agencies the role of coordinating activities in a certain field. Otherwise inter-agency coordination could not be done. The weakness of the inter-agency system, which we are working on now, is that it has focused a lot of its activities to support the CSD. Now everybody wants to start coordination at the national or regional level. The support structure for the global level is more or less ready, now we should take it down to other levels.

Joke Waller Hunter, UN Division for Sustainable Development

of work carried out by other intergovernmental fora. In particular it was felt to be unnecessary for the CSD to address issues such as desertification and biodiversity, which were covered by existing UN conventions and already had functioning review processes.[6]

Delegates therefore approached the future work of the CSD with two main objectives: to narrow its work to areas that complemented and contributed to other processes; and to build on the successful ways of working already established.

Negotiations

The role played by the CSD since its creation was widely applauded. Specific reference is made to the need for continued high-level policy debate to build consensus and catalyse action and commitment to sustainable development. The CSD should work with subsidiary bodies of ECOSOC and others, taking into account follow-up to other related UN conferences. Attention to the challenges of globalization is also called for, and the need for more focused work that avoids duplication of work carried out elsewhere is stressed.

Debate on the proposed programme of work for the CSD (reproduced in Annex 4, page 188) built upon these priorities. The EU proposed that poverty and consumption and production patterns be overriding issues for each year and this was eventually accepted. A framework for the multi-year programme of work, also put forward by the EU, identified a sectoral theme, a cross-sectoral theme and an economic sector or major group to be considered each year. This significantly narrows the range of issues the CSD should consider; these include a number of areas not covered in Agenda 21. Items agreed include the following:

- energy – the CSD will address energy issues in 2001; an intergovernmental group on energy and sustainable development will be set up at the CSD session in 1999;
- forests – the Intergovernmental Forum on Forests will report to the CSD in 1999, this could leave sufficient time for negotiation of a convention before the ten-year review of outcomes from Rio in 2002;
- fresh water – will be considered at the 1998 session, with a view to developing a programme of work to be overseen by the CSD.

New issues on the CSD's agenda are energy, tourism and transport. A full review of progress will take place in 2002.

Commentary

Continuation of the task managers system and the greater focus in the workplan of the CSD have been widely welcomed. It remains to be seen whether the emphasis on areas in which the CSD has the capacity to play a significant role will result in participation by officials from a wider range of government departments, and also from relevant UN bodies and NGOs in greater numbers and with more variety than previously.

It is unclear how issues not addressed in Agenda 21 will be dealt with by the CSD. It may be possible for the secretariat to produce a compilation of references to aspects of the subject from UN conferences and meetings, which could serve as the basis for discussions. This would probably work better for some issues than for others – transport is comprehensively addressed in a wide range of UN documents, whereas tourism has not been considered on anything like the same scale. Alternatively, the secretariat could be given a wider remit to prepare an overview of such subjects, drawing on a much wider range of sources.

4: METHODS OF WORK OF THE COMMISSION ON SUSTAINABLE DEVELOPMENT

Background

The CSD is the body within the UN system that is particularly responsible for the follow-up to the Rio Summit. It has been charged with playing a central role in monitoring and promoting Agenda 21 implementation at every level, from the activities of intergovernmental bodies to the formulation of Local Agenda 21s at community level. Governments are required to report annually to the CSD on implementation of specific parts of Agenda 21. The concept of and criteria for reporting have proved contentious. Even so, 11 countries made presentations on their national reports in 1995 and 1996, opening themselves to comment and criticism in ways that would not have been considered likely at the Rio Summit.

The CSD has also developed new ways in which the UN system can interact with organizations of civil society. The identification of nine 'major groups' of civil society in Agenda 21 has led to active participation by a wide range of organizations in the CSD's work. The CSD itself has come to be seen as a testing ground for new ways of involving NGOs in UN processes.

Negotiations

In considering the future working arrangements for the CSD the following areas were addressed:

GOVERNMENT INPUT

The need for high-level national policy makers to play a more active role in the CSD's work is stressed. Ministers of environment and development should continue to attend, and ways in which the high-level segment of the CSD meeting could be improved should be looked into.

EXCHANGE OF INFORMATION AND EXPERIENCE

The EU put forward text calling for the CSD to oversee a voluntary process of peer review on Agenda 21 implementation, along the lines developed by the OECD. This was strongly resisted by the G77 and others. The final text refers to 'voluntary regional exchange of national experience' and to the possibility of reviews taking place among willing countries on a regional basis. National reporting should continue in streamlined form, taking into account results of the pilot phase on indicators of sustainable development.

LINKS WITH OTHER INTERGOVERNMENTAL BODIES

Closer links with international institutions working on finance, trade and development are called for. The World Bank, Global Environment Facility, UNDP, WTO, UNCTAD and UNEP are also asked to integrate deliberations from the CSD into their activities.

INVOLVEMENT OF MAJOR GROUPS

The Canadian government put forward text calling for more formal steps to involve major groups in the work of the CSD. This was whittled down in negotiations to a call for more of the same dialogue sessions and round tables as have been held over the previous three years. The role of the scientific community is singled out in the final text; a reference to the need for industry to become more accountable was opposed by the United States, who said it would be inappropriate to focus on one major group in this way.

ARRANGEMENTS FOR CSD SESSIONS

Discussion on these issues drew directly on the experience of many of those present. Calls for shorter annual meetings (two weeks maximum) and the identification of key elements for discussion in advance reflect lessons learn from past years. A recommendation to the UN Economic and Social Council that the CSD Bureau should be elected at the end of the annual meeting would allow continuity through the preparations for each annual session, with a high-profile team providing a focus through to the end of the following session. This is at odds with common UN practice, but was felt by many to be an important lesson from the build-up to the Special Session in particular.[7] The Secretary General is invited to review the functioning of the High-Level Advisory Board on Sustainable Development. In his subsequent reform proposals, released in July 1997, Kofi Annan recommended that the board should be disbanded.

RELATIONSHIP WITH OTHER ECOSOC BODIES

ECOSOC is invited to consider ways in which the Committee on New and Renewable Sources of Energy and on Energy for Development and the Committee on Natural Resources should work more closely with the CSD.

Commentary

The use of dialogue and panel sessions during CSD meetings has provided a means for NGOs and governments to engage on more fruitful ground than the negotiation of an agreed text. Within the UN system, the CSD is also thought to have instigated some useful innovations (such as the task managers system) and could lead to others (election of the chair at the end of annual sessions, for instance). The text agreed at the Special Session suggests some areas in which this could be the case. Pressure from the EU for development of a voluntary system of peer review would seem to be the most significant of these.

NOTES

1. No coordinated action in this area as agreed upon, but ECOSOC did decide to carry out an overall review on the theme of poverty eradication in 1999 as a contribution to the Special Session to review outcomes from the Social Summit in 2000.
2. '10 Reasons to be Concerned about the Multilateral Agreement on Investment', Mark Valliantos and Andrea Durbin, FOE, 1997
3. No statement from EOCSOC was agreed on this issue.' The Australian delegate at the ECOSOC session said it was 'a pity and a shame' that the General Assembly document did not refer to the topic, and that the Council had been unable to come up with a reference.
4. The Norwegian and United States governments were unable to persuade others at the ECOSOC that the panel would add to existing arrangements. Norway and the United States pledged to push for such a process in the future.
5. Decision 19/1 of the UNEP Governing Council, reproduced in document A/s-19/5, annex, section I.
6. To some extent the same applies to consideration of climate change. However, some of the factors at the heart of the climate change problem, such as transport and energy use are not comprehensively addressed by other bodies.
7. Approved at the 1997 ECOSOC Session.

Annex 1

Enabling Resolution for the Special Session

UNITED NATIONS GENERAL ASSEMBLY
A/RES/51/181 20 JANUARY 1997
FIFTY-FIRST SESSION AGENDA ITEM 97 (B)

RESOLUTION ADOPTED BY THE GENERAL ASSEMBLY
[ON THE REPORT OF THE SECOND COMMITTEE (A/51/605/ADD.2)]

51/181. SPECIAL SESSION FOR THE PURPOSE OF AN OVERALL REVIEW AND APPRAISAL OF THE IMPLEMENTATION OF AGENDA 21

The General Assembly,

Recalling its resolution 47/190 of 22 December 1992, in which it decided to convene, not later than 1997, a special session for the purpose of an overall review and appraisal of the implementation of Agenda 21,[1]

Reaffirming its resolution 50/113 of 20 December 1995, as the agreed basis that determines the modalities for the preparations for the special session, including the relevant role of the Commission on Sustainable Development, as the functional commission of the Economic and Social Council mandated to follow up the United

Nations Conference on Environment and Development, as well as the role of other relevant organizations and bodies of the United Nations system.

Strongly reaffirming that the special session for the overall review and appraisal of the implementation of Agenda 21 will be undertaken on the basis of and in full respect of the Rio Declaration on Environment and Development,[2]

Taking note of the progress report of the Secretary-General on the state of preparations for the 1997 special session,[3] and taking into account the views and concerns expressed by delegations to the Commission on Sustainable Development at its fourth session, the Economic and Social Council at its substantive session of 1996 and the Second Committee of the General Assembly at its fifty-first session,

1. Decides to convene the special session envisaged in its resolution 47/190 for a duration of one week, from 23 to 27 June 1997, at the highest political level of participation;

2. Decides also that the Commission on Sustainable Development will devote the forthcoming meeting of its Ad Hoc Open-ended Inter-sessional Working Group, to be held from 24 February to 7 March 1997, to preparing for the special session, and that the Commission will devote its fifth session, to be held from 7 to 25 April 1997 as a negotiating meeting, to final preparations for the special session for the overall review and appraisal of the implementation of Agenda 21;

3. Recognizes the important contributions made by major groups, including non-governmental organizations, at the United Nations Conference on Environment and Development and in the implementation of its recommendations, and the need for their effective participation in preparations for the special session, as well as the need to ensure appropriate arrangements, taking into account the practice and experience gained at the Conference, for their substantive contributions to and active involvement in the preparatory meetings and the special session, and in that context invites the President of the General Assembly, in consultation with Member States, to propose to Member States appropriate modalities for the effective involvement of major groups in the special session;

4. Decides to invite States members of the specialized agencies which are not members of the United Nations to participate in the work of the special session in the capacity of observers;

5. Stresses that there should be no attempt to renegotiate Agenda 21, the Rio Declaration on Environment and Development, the Non-legally Binding Authoritative Statement of Principles for a Global Consensus on the Management, Conservation and Sustainable Development of All Types of Forests[4] or other internationally recognized intergovernmental agreements in the field of environment and sustainable development, and that discussions at both the preparatory meetings and the special session should focus on the fulfilment of commitments and the further implementation of Agenda 21 and related post-Conference outcomes;

6. Requests the Secretariat to provide all relevant reports called for in General Assembly resolution 50/113, including all other reports related to the outcome of the United Nations Conference on Environment and Development, for consideration by the Ad Hoc Open-ended Inter-sessional Working Group of the Commission on Sustainable Development and by the Commission at its fifth session, in accordance with the six-week rule and preferably not later than 15 January 1997;

7. Requests the Secretary-General to ensure that preparations for the comprehensive report are conducted in accordance with paragraph 13 (a), (b), (c) and (d) of Assembly resolution 50/113;

8. Invites the Secretary-General to include in the reports requested in Assembly resolution 50/113 for the preparation of the special session information on the application of the principles contained in the Rio Declaration, and invites the Governing Council of the United Nations Environment Programme to include in its report to the General Assembly at its special session information and views on ways to address, in a forward-looking manner, national, regional and international application of these principles and the implementation of Agenda 21 in the interrelated issues of environment and development;

9. Decides to consider at its special session, inter alia, the application of the principles of the Rio Declaration at all levels – national, regional and international – and to make relevant recommendations thereon;

10. Requests that other contributions to the special session, in addition to those identified in its resolution 50/113, include submissions from relevant bodies and organizations of the United Nations, including the Ad Hoc Intergovernmental Panel on Forests of the Commission on Sustainable Development and the Global

Environmental Facility, information on the outcomes of United
Nations conferences held since the United Nations Conference on
Environment and Development, such as the Programme of Action
for the Sustainable Development of Small Island Developing
States,[5] regional and subregional conferences, summits and other
inter-sessional meetings on sustainable development organized by
countries, and information on the activities of relevant United
Nations conventions on the environment and development and
the global freshwater assessment, and that account be taken of the
activities organized by major groups, including business and
industry and non-governmental organizations;

11. Requests the Secretary-General, in the report on cross-sectoral
    issues of Agenda 21 for the special session, to give special consid-
    eration, without prejudice to other priority issues that may be
    identified in the preparatory process, to combating poverty and
    to health, financial resources and mechanisms, education,
    science, transfer of technology, consumption and production
    patterns, trade, environment and sustainable development, major
    groups, demographic dynamics, capacity-building and
    decision-making;

12. Also requests the Secretary-General, in the reports for the special
    session, to give consideration, where appropriate and without
    prejudice to other priority issues that may be identified in the
    preparatory process, to linkages between the cross-sectoral issues
    of Agenda 21 and relevant sectoral issues;

13. Welcomes the outcome of the United Nations Conference on
    Human Settlements (Habitat II), held at Istanbul from 3 to 14
    June 1996, and its relevance to the field of sustainable develop-
    ment, calls for effective interaction and exchange of information
    on work carried out by the Commission on Sustainable
    Development and the Commission on Human Settlements, and
    invites the Commission on Human Settlements to make a contri-
    bution to the special session in connection with the
    implementation of the Habitat Agenda[6] adopted in Istanbul;

14. Invites Governments and regional organizations to cooperate
    with the Secretary-General in preparing country profiles for
    review at the fifth session of the Commission on Sustainable
    Development, as envisaged in paragraph 13 of General Assembly
    resolution 50/113;

15. Also invites Governments to assist developing countries, particu-
    larly the least developed among them, in participating fully in the

special session and its preparatory process, and to make timely contributions to the Trust Fund for Support of the Work of the Commission on Sustainable Development;

16. Requests the Secretary-General to enhance the public information programme of the United Nations so as to raise global awareness in a balanced manner, in all countries, of both the special session and the work undertaken by the United Nations in the follow-up to the Conference, and invites all Governments to promote widespread dissemination, at all levels, of the Rio Declaration on Environment and Development, and to make voluntary contributions to support the public outreach activities of the United Nations for the special session;

17. Decides to include in the provisional agenda of its fifty-second session the sub-item entitled "Special session for the purpose of an overall review and appraisal of the implementation of Agenda 21", and requests the Secretary-General to submit to it at that session a report on the special session.

86th plenary meeting 16 December 1996

NOTES

1 Report of the United Nations Conference on Environment and Development, Rio de Janeiro, 3–14 June 1992 (A/CONF.151/26/Rev.1 (Vol. I and Vol.I/Corr.1, Vol. II, Vol. III and Vol. III/Corr.1)) (United Nations publication, Sales No. E.93.I.8 and corrigenda), vol. I: Resolutions Adopted by the Conference, resolution 1, annex II.

2 Ibid., annex I.

3 A/51/420.

4 Report of the United Nations Conference on Environment and Development, Rio de Janeiro, 3–14 June 1992 (A/CONF.151/26/Rev.1 (Vol. I and Vol. I/Corr.1, Vol. II, Vol. III and Vol. III/Corr.1)) (United Nations publication, Sales No. E.93.I.8 and corrigenda), vol. I: Resolutions Adopted by the Conference, resolution 1, annex III.

5 Report of the Global Conference on the Sustainable Development of Small Island Developing States, Bridgetown, Barbados, 25 April–6 May 1994 (United Nations publication, Sales No. E.94.I.18 and corrigendum), chap. I, resolution 1, annex II.

6 A/CONF.165/14, chap. I, resolution 1, annex II.

Annex 2

Organization of the Special Session

UNITED NATIONS A/S–19/2 12 June 1997
19th General Assembly Nineteenth special session Item
6 of the provisional agenda A/S–19/1.

ORGANIZATION OF THE SESSION

Note by the President of the General Assembly

1. In accordance with General Assembly resolutions 47/190 of 22
December 1992, 50/113 of 20 December 1995 and 51/181 of 16
December 1996, the nineteenth special session of the General
Assembly for the purpose of an overall review and appraisal of
the implementation of Agenda 21 was convened for a duration of
one week, from 23 to 27 June 1997. The provisional agenda of
the special session was issued as document A/S–19/1.

2. Taking into consideration the practice of previous special
sessions, the relevant provisions of General Assembly resolution
51/181 and decision 51/467 of 18 April 1997, and the informal
consultations held by the President of the General Assembly at its
fifty-first session with Member States on 2 and 4 April 1997 on
the organization of the special session, the Assembly may wish to
consider the following arrangements for the nineteenth special
session:

1. President

3. The nineteenth special session shall take place under the presidency of the President of the fifty-first regular session of the General Assembly.

2. Vice-Presidents

4. The Vice-Presidents of the nineteenth special session shall be the same as at the fifty-first regular session of the General Assembly.

3. General Committee

5. The General Committee of the nineteenth special session shall consist of the President of the special session, the 21 Vice-Presidents of the special session, the Chairmen of the six Main Committees of the fifty-first regular session of the General Assembly and the Chairman of the Ad Hoc Committee of the Whole of the special session.

4. Credentials Committee

6. The Credentials Committee of the nineteenth special session shall have the same membership as the Credentials Committee of the fifty-first regular session of the General Assembly.

5. Ad Hoc Committee of the Whole

7. The nineteenth special session shall establish an Ad Hoc Committee of the Whole of the nineteenth special session. The Bureau of the Ad Hoc Committee of the Whole shall consist of one Chairman, three Vice-Chairmen and one Rapporteur.

6. Rules of procedure

8. The rules of procedure of the General Assembly shall apply at the nineteenth special session.

7. Level of representation

9. In paragraph 1 of resolution 51/181, the General Assembly decided that representation should be 'at the highest political level of participation'.

8. Allocation of items

10. The substantive item (overall review and appraisal of the implementation of Agenda 21) shall be allocated to the Ad Hoc

Committee of the Whole of the nineteenth special session for
consideration, on the understanding that the debate on the item
shall take place in plenary meeting. The Ad Hoc Committee would
be entrusted with the task of considering all proposals submitted
and preparing the final document(s) for consideration and
adoption by the General Assembly. The Ad Hoc Committee of the
Whole will also hear statements from observers, representatives of
United Nations programmes, specialized agencies and others.

9. Debate on the overall review and appraisal of the implementation of Agenda 21

11. The debate on the overall review and appraisal of the implemen-
    tation of Agenda 21 shall begin in plenary meeting on the
    morning of Monday, 23 June and end in the afternoon of Friday,
    27 June 1997. Owing to the large number of speakers expected
    to take the floor in a limited amount of time, statements should
    not exceed 7 minutes.
12. Without creating a precedent, Heads of State, Vice-Presidents,
    Crown Princes and Princesses and Heads of Government shall
    have equal standing as concerns the list of speakers.

10. Participation of speakers other than Member States in the debate

13. States members of the specialized agencies which are not
    members of the United Nations but which were invited to partici-
    pate in the work of the special session in the capacity of
    observers in paragraph 4 of General Assembly resolution 51/181,
    and intergovernmental and other organizations and entities
    having received a standing invitation to participate as observers
    in the work of the Assembly will be invited to participate in the
    debate in plenary meeting. The Ad Hoc Committee of the Whole
    will also hear statements by observers.
14. Representatives of United Nations programmes, specialized
    agencies and others will be invited to participate in the debate in
    plenary meeting, provided they are at the highest level. The Ad
    Hoc Committee of the Whole will also hear statements by repre-
    sentatives of United Nations programmes, specialized agencies
    and others.
15. In accordance with General Assembly decision 51/467 of 18 April
    1997, the President of the Assembly extended invitations to
    representatives of major groups, as identified in Agenda 21 and

represented by non-governmental organizations in consultative status with the Economic and Social Council and on the roster, to participate in the debate in plenary meeting.

Annex 3

E/CN.17/1997/Z 31 January 1997
Overall Progress Achieved since the United Nations Conference on Environment and Development

REPORT OF THE UN SECRETARY-GENERAL

CONTENTS

INTRODUCTION

1. The present report was prepared in accordance with the General
Assembly resolution 50/113 for the fifth session of the
Commission on Sustainable Development, which will be devoted
to preparations for the special session of the General Assembly to
be held from 23 to 27 June 1997 for the purpose of an overall
review and appraisal of the implementation of Agenda 21. The
report also takes into account relevant provisions of General
Assembly resolution 51/181 concerning the special session.

2. This report contains a global assessment of the current status of
economic and social development and environmental sustainability, followed by an appraisal of progress made since the United
Nations Conference on Environment and Development
(UNCED), focusing on the main achievements and unrealized
expectations. It also attempts to identify the main challenges and
priorities in the implementation of Agenda 21 and of other
outcomes of UNCED for the period after the 1997 review, including the future role of the Commission on Sustainable
Development. The structure of the report takes into account the
three main interrelated components of sustainable development,
namely economic growth, social development and environmental
sustainability.

3. The report does not analyse progress or lack thereof in the
implementation of individual chapters of Agenda 21, nor does it
describe all relevant activities or policy changes undertaken as a
follow-up to UNCED at the international, regional and national
levels or by the major groups. More detailed information on

these actions can be found in the addenda to this report
(E/CN.17/1997/2/Add.1–30) as well as in document
E/CN.17/1997/5, which assesses progress achieved at the national
level on the basis of information contained in the 'country
profiles' prepared in cooperation with the Governments
concerned.

4. A number of conclusions presented in the report are based on
the information contained in other reports prepared for the 1997
review, in particular in the reports on critical trends in sustainable development (E/CN.17/1997/3), the results of the
comprehensive freshwater assessment (E/CN.17/1997/9) and
activities that pose a major threat to the environment
(E/CN.17/1997/4). Furthermore, in the preparation of the report
account was taken of the outcomes of other recent international
conferences, as well as of other major studies and reports that
deal with issues relevant to sustainable development and the
implementation of Agenda 21.

I. GLOBAL ASSESSMENT

A. SUSTAINABLE DEVELOPMENT IN THE YEARS SINCE THE UNITED NATIONS CONFERENCE ON ENVIRONMENT AND DEVELOPMENT

5. At the United Nations Conference on Environment and
Development, held at Rio de Janeiro in 1992, Governments
adopted Agenda 21, a programme of action for sustainable devel-
opment worldwide,[1] the Rio Declaration on Environment and
Development[2] and the Non-legally Binding Authoritative
Statement of Principles for a Global Consensus on the
Management, Conservation and Sustainable Development of All
Types of Forests (the Forest Principles).[3] Sustainable development may be regarded as the progressive and balanced
achievement of sustained economic development, improved
social equity and environmental sustainability. Accordingly,
Agenda 21 stresses the importance of integrated policy development, citizen participation in decision-making including the full
participation of women, institutional capacity-building and global
partnerships involving many stakeholders.

6. Sustainable development is about change: change in development paths; change in the production and consumption patterns that determine how the needs – and often wants – of people are met and, in turn, contribute to – or hinder – development. It is evident that all countries must have the opportunity to realize economic growth in order to meet their essential needs. But the quality of growth is as important as its quantity. Is it promoting equity? Does it contribute to meeting the food, health care, safe water, shelter and educational needs of the developing countries, in particular the least developed countries? Does it lead to an environment which is conducive to a healthy and productive life, as advocated in principle 1 of the Rio Declaration? Does it take a precautionary approach to exploitation of the planet's ecosystems? Economic development, social development and environmental protection are mutually reinforcing components of sustainable development. The importance given to each of the components may vary from country to country.

7. The series of global conferences organized by the United Nations in the years since UNCED have all incorporated the fundamental principles and objectives of Agenda 21. The analysis and the plans of action emerging from these conferences are essential to our understanding of sustainable development and to ultimately achieving it.

8. The five years since UNCED have been characterized by accelerated "globalization", which refers to the growing interaction of countries in world trade, foreign direct investment and capital markets. The globalization process has been abetted by technological advances in transport and communications and by a rapid liberalization and deregulation of trade and capital flows, at both the national and international levels. Democracy has continued to spread and become more consolidated in countries where democratic forms of government have only recently been established. The end of the cold war has permitted an overall reduction in military expenditures as a share of gross domestic product (GDP). Yet preoccupation with fiscal consolidation in many developed market economies has led to some shrinkage of social safety nets and stagnation, if not outright reductions, in the volume of official development assistance (ODA). At the same time, regional conflicts, communal strife and civil war have visited unspeakable tragedy on millions of people – the very antithesis of sustainable development.

B. ECONOMIC GROWTH

1. ECONOMIC PERFORMANCE

9. During the period 1992–1996 growth of GDP in the developing countries averaged about 5.3 per cent per year, compared with 3.1 per cent during the 1980s and 4.2 per cent during the period 1991–1992. This acceleration of GDP growth has permitted per capita GDP to rise by more than 3 per cent per year during the past four years; moreover, growth gradually spread as the number of countries exhibiting increasing per capita GDP rose from an average of 55 countries (containing about 83 per cent of the population living in developing countries) during the years 1990–1993 to 75 countries (accounting for 96 per cent of the population) in 1996. This pattern was not, however, shared by sub-Saharan African countries and least developed countries where per capita GDP continued to fall or stagnate through 1995. Worldwide, unequal income distribution within countries means that over 1.5 billion people did not share in economic growth and experienced declining per capita incomes in the 1990s.

10. This improved growth performance was due more to successful national policies than to external circumstances. Growth of world output was considerably slower during the first half of the 1990s compared with the decade of the 1980s. Consequently, during the period 1991–1993 growth in world trade was relatively slow and real commodity terms of trade continued the decline that began in 1989. These trends were reversed during the period 1994–1996 with the resumption of sustained growth in world output. Developing countries' export volumes increased more rapidly than world trade throughout the 1990s, averaging about 12 per cent per year during the period 1994–1996. Export volume growth was exceptional in Latin America, South and East Asia and China. The strengthening of regional trading arrangements was an important factor in explaining the export growth performance of developing countries as well as their continued penetration of markets for manufactured goods in developed market economies. Accompanying the GDP and export performance of developing countries and China, manufacturing value added has grown as well, increasing its share in world totals from about 15 per cent in 1991 to about 18 per cent in 1995.

2. ENERGY AND MATERIAL USE

11. Economic growth and social development depend on energy use. World energy consumption has risen steadily and by 1993 was more than 40 per cent higher than in 1973. Global demand for energy continues to rise to meet the needs of a growing world population. The growth in per capita energy demand will continue because an increase in per capita energy is linked to the growth of the world economy, particularly for developing economies. Major increases in energy-generating capacity are still required in many developing countries if basic human needs are to be met. Over 2 billion people still have little or no access to public and/or commercial energy supplies.

12. Consumption of some materials is stabilizing in the industrialized countries, as a result of improved efficiency and economic restructuring, but consumption is rising rapidly in developing economies. However, reflecting the very large differences in per capita income between regions, per capita consumption of commercial materials remains far higher in developed countries.

3. DEVELOPMENT FINANCE

13. Trends in international capital flows have been mixed. Net capital flows of private direct investment, portfolio investment and commercial bank lending have increased during the period 1992–1995, but have been concentrated in a relatively small number of developing countries. Net flows of official development assistance, on which least developed and several other low-income countries depend, after increasing in 1993, declined in real terms in both 1994 and 1995. These trends in ODA are clearly disappointing when viewed against the expectation at UNCED for new and additional net flows. Perhaps reflecting heightened concern for social development and environmental management, there has been, however, a small but noticeable increase in the proportion of official development finance flowing into these areas.

C. SOCIAL DEVELOPMENT

14. Data on social development, other than demographic data, are still not monitored as frequently or as comprehensively as

economic data. For the most part, there is little data more recent than for 1993. Such data as are available generally show a small but positive change in a number of social indicators, and that this progress is evident in most world regions. In sub-Saharan Africa, however, a number of indicators show worsening trends during the 1990s. In many countries in transition also, several social indicators show worsening trends.

1. DEMOGRAPHIC DYNAMICS

15. According to the United Nations 1996 Revision[4] of global population and demographic estimates and projections, population calculations for the period 1990–1995 indicate that growth fell faster, national fertility declines were broader and deeper and migration flows larger than previous estimates had indicated. The latest medium fertility variant projection shows that the world population will stabilize at about 9.4 billion in 2050, almost half a billion lower than the figure projected in the 1994 Revision.[5]

2. HEALTH

16. One measure of global health – life expectancy – increased slightly between 1985–1990 and 1990–1995. Increases were registered in most countries but 15 countries in sub-Saharan Africa and 17 countries with economies in transition experienced declines in life expectancy. Another important health indicator, per capita dietary energy supply (kilocalories), shows that in 1990–1992 as compared with the period 1979–1981, increases occurred in nearly all world regions. The exceptions were Eastern Europe and sub-Saharan Africa. Increases were largest in South, East and South-East Asia. However, the increase was negligible in Latin America. Estimates of the prevalence of underweight children display similar patterns. Eight hundred and forty million people in the world suffer from malnutrition. A number of infectious diseases may be eradicated in the near future, given continued efforts, but others, notably malaria, are increasing.

17. Excessive environmental pollution is affecting the health of millions of people in urban agglomerations in developing countries. While the developing countries as a whole have narrowed the "health gap" with the industrialized countries in several important indicators, including life expectancy and infant and child mortality, the gap is widening between the least developed countries and other developing countries.

3. DRINKING WATER AND SANITATION

18. In spite of efforts since the start of the International Drinking Water Supply and Sanitation Decade in 1981, some 20 per cent of the world's population lacks access to safe water and 50 per cent lacks access to safe sanitation. At any given time, an estimated 50 per cent of the population in developing countries is suffering from water-related diseases caused either directly by infection, or indirectly by disease-carrying organisms. The World Health Organization estimates that more than 5 million people die each year from diseases caused by unsafe drinking water and a lack of sanitation. In terms of the economic impact of poor water supply systems, it has been estimated that the provision of safe water supply, suitably located, could save over 10 million person-years of efforts in fetching water, mostly by women and female children, in developing countries. Water supplies in many cities of developing countries are intermittent, and an increasing number of peri-urban poor are without services and are often left at the mercy of private vendors charging exorbitant prices.

4. EDUCATION

19. Indicators of education such as enrolment ratios and adult literacy show improvement in all developing country regions. Between 1990 and 1993 combined gross enrolment ratios for all levels of education increased, with those of females increasing slightly more than those of males. However, increases were very slight in sub-Saharan Africa and in the least developed countries, groupings where gross enrolment ratios are very much lower than in other regions. Rates of illiteracy among adults appear to have declined steadily in all developing regions, including South Asia and sub-Saharan Africa; illiteracy rates for all developing countries averaged 29 per cent in 1993.

5. POVERTY

20. Both economic growth and investment in human resources have impacts on the incidence of poverty; moreover, environmental degradation and poverty can interact in a vicious circle. Data on the incidence of poverty indicate that the percentage of those living in poverty in developing countries declined slightly between 1990 and 1993 but all of the improvement was concentrated in East Asia and the Pacific, where the absolute number of

poor declined as well. In other developing country regions the
number of poor actually increased; globally, the numbers of
people living in absolute poverty rose to 1.3 billion in 1993.
Women continue to be disproportionately affected. In rural areas,
the number of women in absolute poverty rose by nearly 50 per
cent in the past two decades and they now constitute a substantial majority of the world's poor.

D. ENVIRONMENTAL SUSTAINABILITY

21. Services provided by the environment are essential for economic
activity, human health and the preservation of life itself.
Imprudent depletion or degradation of natural resources or
exceeding the capacity of air, soil and water to absorb pollutants
will undermine long-term prospects for economic growth just as
surely as will failure to maintain and increase stocks of physical
capital or failure to invest in human development.

1. IMPACTS OF ENERGY USE

22. Current forms of energy production and use, primarily based on
fossil fuels, have serious adverse effects on the environment:
emissions contaminate air, water and soil and contribute to
global warming. Developed market economies have achieved a
significant reduction in energy intensity due to improvements in
generation and end-use efficiency in many socio-economic
sectors. However, the increased volume of economic activity has
offset these gains, and emissions of carbon dioxide continue to
rise. Experiences in developing countries have varied considerably, even among those countries within the same region,
because of the significant differences in their resources base,
energy demand structures, economic situation, technological
capacity and population and development strategy.

23. Most developed countries, and a number of middle-income
developing countries, have experienced significant reductions in
some other energy-related emissions, notably sulphur dioxide.
The resulting improvements in local air and water quality can be
attributed both to technological change responding to the operation of market forces, and to increasingly stringent regulation of
ambient quality standards and emissions, especially from motor
vehicles.

2. FRESHWATER

24. The analysis carried out under the Comprehensive Assessment of the Freshwater Resources of the World gives rise to serious concerns as to the sustainability of current pathways of water resources development and utilization in many developed and developing countries alike. Global demand for water has increased dramatically over the past century and it is estimated that more than 8 per cent of the world's population now lives in countries that are highly water stressed and another 25 per cent in countries that are experiencing moderate to high water stress. If current trends in water use persist, two thirds of the world's population could be living in countries experiencing moderate or high water stress by 2025.

25. The present situation and current trends have serious implications in terms of economic development and food production. Unless managed with a view to achieving much greater efficiency, for which there is considerable potential, water resources could become a serious factor limiting socio-economic development in many developing countries. Efforts to increase efficiency and maximize economic benefits will tend to shift water users away from low value products. This, in turn, will have a serious impact on poor farmers operating inefficient irrigation schemes unless policies are designed and implemented to mitigate the impact of such a shift.

26. In addition, a number of developed countries are facing stressful conditions with regard to the utilization of their water resources, in many cases due to deteriorating quality. Freshwater reserves continue to be used as a sink for the dumping of wastes from urban and industrial sources, agricultural chemicals and other human activities. Current estimates indicate, for instance, that about 90 per cent of waste waters from urban areas are discharged without any treatment in many developing countries. Altogether, water quality has degraded rapidly, and in some regions has become so bad that groundwater is not suitable even for industrial use.

3. SOIL QUALITY AND FOOD PRODUCTION

27. The most recent comprehensive survey of soil degradation (GLASOD) indicates that faulty agricultural practices are a significant cause of soil degradation; examples include nutrient mining

due to cropping intensification, insufficient fertilizer inputs, erosion and overgrazing by livestock. As much as 10 per cent of the earth's vegetated surface is now at least moderately degraded. Continued degradation of the agricultural land base will have serious implications for future food security at the local level.

28. It is expected that about two thirds of the increases in agricultural production required to meet projected increases in effective demand will come as a result of improved yields from land currently under cultivation in developing countries, much of which is irrigated. In Africa and Latin America and the Caribbean, increases in yields are expected to contribute upwards of 50 per cent of production increases. An additional 21 per cent of the increase in production is expected to be achieved through a projected expansion of harvested areas, particularly in sub-Saharan Africa, Latin America and East Asia. Of the projected 124 million hectares in new harvested areas, new irrigated lands are viewed as being limited to 45 million hectares. Increases in cropping intensity are expected to contribute the remaining 13 per cent of the total increases in food production.

29. While the medium-term prospects for increasing food production are good, trends in soil quality and the management of irrigated land raise serious questions about longer-term sustainability. It is estimated that about 20 per cent of the world's 250 million hectares of irrigated land are already degraded to the point where crop production is significantly reduced.

4. FOREST COVER

30. According to the Food and Agriculture Organization of the United Nations (FAO) in the *State of the World's Forests, 1995*, deforestation and degradation remain the major issues. For the period 1980–1990, the annual estimated loss in natural forest area is 12.1 million hectares. Disaggregated estimates are as follows: global change in forests and other wooded lands was –10.0 million hectares per year; natural forest loss in developing countries, 16.3 million hectares per year; increase in plantations in developing countries, 3.2 million hectares per year. The rates and causes of deforestation differ greatly between countries and regions; determining factors include population density and growth rates, levels and rates of development, the structure of property rights and cultural systems. Rates of tropical deforestation increased in each of the past three decades in all tropical

regions and are currently highest in Asia. There is increasing concern about the decline in forest quality associated with intensive use of forests and unregulated access.

31. The largest losses of forest area are taking place in the tropical moist deciduous forests, the zone best suited to human settlement and agriculture; recent estimates suggest that nearly two thirds of tropical deforestation worldwide is due to farmers clearing land for agriculture. A growing proportion of commercial wood consumption needs in developing countries appears to be coming from plantations which, when well managed, are proving highly productive. In temperate developing countries, increases in plantation forest exceeded the declines in natural forests. There appears to have been a net gain in area of forest and other wooded land in most regions.

5. THE MARINE ENVIRONMENT AND FISHERIES

32. Coastal ecosystems, including wetlands, tidal flats, saltwater marshes, mangrove swamps coastal nursery areas and the flora and fauna that depend on them, are particularly at risk from industrial pollution and urban land conversion. Coastal urban centres are already home to approximately 1 billion people worldwide and are experiencing unprecedented growth, much of it in developing countries. According to a recent study undertaken by the World Resources Institute, about half the world's coasts are threatened by development-related activities. The harmful effects of coastal degradation are often felt first by subsistence fishers and small-scale fleets which operate close to shore. Wider impacts include intensified coastal erosion, decreased protection from storm damage and loss of biodiversity.

33. The marine fisheries account for about 82 per cent of the total global fish harvest. The marine harvest has continued to increase slowly since 1970 despite a small decrease in early 1990; the additional production is coming mainly from highly fluctuating small pelagic wild resources and marine and coastal aquaculture. According to a new assessment made by FAO in late 1996, 25 per cent of the world's marine fisheries are being fished at their level of maximum productivity and 35 per cent are overfished (yields are declining). In order to maintain current per capita consumption of fish, global fish harvests (110 million tons in 1994) must be increased; FAO estimates that much of the increase must come from mainly inland aquaculture. This expansion is not without

risk, since aquaculture is a known source of water pollution, wetland loss and mangrove swamp destruction. Expansion will also be constrained by land-based pollution.

6. BIODIVERSITY

34. Biodiversity is increasingly under threat from development, which destroys or degrades natural habitats, and pollution from a variety of sources. The first comprehensive global assessment of biodiversity was released in 1995 at the second meeting of the Conference of the Parties to the Convention on Biological Diversity.[6] It put the total number of species at close to 14 million and found that between 1 and 11 per cent of the world's species per decade may be threatened by extinction. Major threats to species are related to threats to the ecosystems that support them from both development and pollution. There is, thus, a direct link to the forest agenda. Coastal ecosystems, which host a very large proportion of marine species, are at great risk, with perhaps one third of the world's coasts at high potential risk of degradation and another 17 per cent at moderate risk. FAO estimates that the rural poor in developing countries depend upon biological resources to meet about 90 per cent of their needs; the social and economic value of biodiversity is thus very high.

7. WASTE AND HAZARDOUS MATERIALS

35. Domestic and industrial waste production continues to increase in both absolute and per capita terms, worldwide. In the developed world, per capita waste generation has increased threefold over the past 20 years; in developing countries, it is highly likely that waste generation will double during the next decade. The level of awareness regarding the health and environmental impacts of inadequate waste disposal remains rather poor; poor sanitation and waste management infrastructure is still one of the principal causes of death and disability for the urban poor.

36. Toxification from dissipative use of modern materials has emerged as an issue of concern. Approximately 100,000 chemicals are now in commercial use and their potential impacts on human health and ecological function represent largely unknown risks. Persistent organic pollutants are now so widely distributed by air and ocean currents that they are found in the tissues of people and wildlife everywhere; they are of particular concern because of their high levels of toxicity and persistence in the

environment. Pollution from heavy metals, especially from their use in industry and mining, is also creating serious health consequences in many parts of the world. Incidents and accidents involving uncontrolled radioactive sources continue to increase. Particular risks are posed by the legacy of contaminated areas left from military activities involving nuclear materials.

II: APPRAISAL OF PROGRESS ACHIEVED SINCE THE UNITED NATIONS CONFERENCE ON ENVIRONMENT AND DEVELOPMENT

37. Since 1992, sustainable development has been more widely accepted as an integrating concept that seeks to unify and bring together economic, social and environmental issues in a participatory process of decision-making. The years since UNCED have seen a growing consensus on the need for integrated approaches, as advocated in Agenda 21, and real progress has been made in establishing a conceptual framework within which planning for sustainable development can take place. Five years after the Conference, it is clear that the policy process is far more advanced in some areas than others. Some are still at the stage of defining problems and agreeing on necessary responses. Others have moved to the stage of target-setting and deployment of new policy instruments to achieve change. In a few cases, intervention has already brought measurable results.

38. The present section discusses the development of integrated strategies for sustainable development at various levels of government and in the outcomes of major international conferences. It then evaluates recent changes in international trade regimes and assesses progress towards changing production and consumption patterns, a key strategic approach to achieving sustainable development identified in Agenda 21. Lastly, it evaluates progress in the management of natural resources, the involvement of various actors and means of implementation.

A. DEVELOPING STRATEGIES FOR SUSTAINABLE DEVELOPMENT

39. Following UNCED, a number of related global plans and strategies have been agreed which attempt to translate the principles

of Agenda 21 into practice. An important example is the
Programme of Action adopted at the Global Conference on the
Sustainable Development of Small Island Developing States
(Bridgetown, Barbados, 1994).[7] The Programme of Action identi-
fies a number of important priority areas for sustainable
development in small island developing States (see
E/CN.17/1997/14).

40. At the regional level, the past five years have seen a number of
initiatives to formulate regional sustainable development strategies or action plans and to establish mechanisms for regional
cooperation in implementing such initiatives. They have often
been launched as the result of regional summits or ministerial
meetings and are intended to translate global issues into regional
ones. An example is the Regional Action Programme on
Environmentally Sound and Sustainable Development in Asia and
the Pacific, 1996–2000. In addition, integrated plans have been
developed and adopted for smaller regions that share or feel
responsible for common resources or ecosystems. Examples are
the plans for the Arctic, the Baltic Sea and, recently initiated, the
Sustainable Development Plan for the Mediterranean.

41. At the national level, countries ranging from China, which
launched the first national Agenda 21, to Swaziland and the
United Kingdom of Great Britain and Northern Ireland, have
produced national sustainable development strategies, national
conservation strategies or environmental action plans.
Developing countries have made particular progress in this
regard. In some cases, this has been done with the assistance of
the World Bank, the United Nations Development Programme
(UNDP), the International Union for Conservation of Nature and
Natural Resources (IUCN) and some bilateral donors. Over 40
African countries have some sort of coordinating mechanism to
produce such plans. Environmental factors have also been incorporated into macroeconomic strategy.

42. The Development Assistance Committee of the Organisation for
Economic Cooperation and Development (OECD) has recognized
the strategic importance of sustainable development strategies. It
has set the target that by the year 2005 national strategies for
sustainable development are to be adopted in all countries, so as
to ensure that current trends in the loss of environmental
resources are effectively reversed at both the global and national
levels by the year 2015.

43. Further progress in the effort to develop sustainable development strategies must take account of such constraining factors as: (i) Governments, particularly of developing countries, are overloaded with requests for various types of strategies, plans and schemes to satisfy the requirements of international banks, lending agencies and international organizations, which have not been adequately coordinated or prioritized; (ii) not all governing bodies of international organizations, even within the United Nations system, have the same understanding of the concept of sustainable development – some have adopted programmes of environmentally sustainable development, others have called for sustainable human development, while others have talked of conservation or other types of environmental plans; this has led to some confusion regarding the core issues of sustainable development; (iii) international agreements are often being reached faster than countries can respond effectively to agreed requirements; (iv) Governments frequently lack the financial and staff resources to implement the different international conferences, conventions and agreements they have agreed to or signed; (v) capacity-building efforts should not stop after sustainable development strategies have been formulated, since the implementation of such strategies requires continuous support.

44. At the local level there has been a positive trend in the number of cities around the world which have formulated and are implementing local Agenda 21s. Currently almost 2,000 local governments from 49 countries are pursuing local Agenda 21 action plans through official planning processes in partnership with the voluntary and private sectors in their communities. The "sustainable cities" process, which started in 1992, has been boosted by the United Nations Conference on Human Settlements (Habitat II).

B. GLOBAL CONFERENCES

45. Since 1992, a number of major United Nations conferences have made policy advances and strengthened commitments to social aspects of sustainable development. The International Conference on Population and Development (Cairo, 1994) emphasized the importance of broad-based economic and social development, including expanded education, health care and

economic opportunities, especially for women, in reducing
desired family size and thereby reducing population growth. The
World Summit for Social Development (Copenhagen, 1995)
emphasized that social development requires not just economic
growth, but also the eradication of poverty, full employment and
social integration. The Fourth World Conference on Women
(Beijing, 1995) pointed out that many activities crucial to sustainable development are largely the responsibility of women, that
women disproportionately bear many of the burdens of environmental degradation, but that women remain largely absent at all
levels of policy formulation and decision-making and that unless
women's contribution to environmental management is recognized and supported, sustainable development will remain an
elusive goal. The United Nations Conference on Trade and
Development (UNCTAD), at its ninth session (Midrand, 1996),
highlighted the potential benefits of globalization and trade
liberalization to developing countries but warned of the risks of
marginalization of poorer countries unable to capitalize on new
opportunities. Habitat II (Istanbul, 1996) raised global awareness
of the key role of human settlements in sustainable development
as the majority of the global population will be living in cities in
the next century, thus increasing the urgency of facing the
growing social, economic and environmental problems of cities.
Finally, the World Food Summit (Rome, 1996) called for renewed
effort to combat hunger, which persists in the poorer regions of
the world and is likely to increase in spite of food surpluses at
the global level. All of these conferences adopted plans of action
that complement Agenda 21, superseding it in some respects. A
number of bodies within and outside the United Nations system,
notably the Economic and Social Council, are coordinating the
implementation of those plans of action.

C. INTERNATIONAL TRADE, ECONOMIC GROWTH AND SUSTAINABLE DEVELOPMENT

46. Globalization and liberalization have increased the potential for
    international trade to become an unprecedented engine of
    growth and an important mechanism for integrating countries
    into the global economy. A good number of developing countries
    have seized the opportunities and seen the rapid growth of their

economies. Not all countries, however, have been in a position to seize these new trading opportunities. There is a real risk that these countries, especially the least developed among them, and other structurally weak economies, could become further marginalized. At the same time, it is widely recognized that the integration and fuller participation of these and other developing countries and countries in transition in the global economy would contribute substantially to the expansion of world trade, serving the overall objectives of world economic growth in the context of sustainable development.

47. The completion of the Uruguay Round of multilateral trade negotiations was a major step by the international community towards the expansion of the rule-based international trading system and advancing liberalization in international trade and creating a more secure trading environment. The Uruguay Round furthered and consolidated the process of trade liberalization through improvements in market access and more stringent disciplines over trade measures. It set out a system of multilateral trade obligations subject to a common dispute settlement mechanism which will place most countries at virtually the same level of multilateral obligation within a relatively short time. Most of the multilateral trade agreements contain their own built-in agenda for review, possible revision and negotiation of future commitments.

48. It has been recognized that, during the reform programme leading to greater liberalization of trade in agriculture, least developed and net food-importing developing countries may experience negative effects in terms of the availability of adequate supplies of basic foodstuffs from external sources on reasonable terms and conditions, including short-term difficulties in financing normal levels of commercial imports of basic foodstuffs. The plight of the least developed countries and the need to ensure their effective participation in the world trading system is also recognized.

49. The least developed countries, particularly those in Africa, and other developing countries remain constrained by weak supply capabilities and are unable to benefit from trade. Marginalization, both among and within countries, has been exacerbated. Too many people continue to live in dire poverty.

50. Intergovernmental deliberations of UNCTAD, the World Trade Organization (WTO), the Commission on Sustainable Development and other international organizations have resulted in a better understanding of the relationship between trade,

environment and development. The post-UNCED debate has
focused on, and will continue to explore, the scope of the
complementarities between trade liberalization, economic development and environmental protection. Governments have taken
appropriate steps to ensure that trade and environment are now
firmly incorporated into the work programmes of WTO, UNCTAD
and other relevant international organizations. The breadth and
complexity of the issues covered in the work programme of the
Committee on Trade and Environment (CTE) of WTO shows that
further work needs to be undertaken and ministers have directed
CTE to continue work on all items of its agenda, as contained in
its report, building on the work accomplished thus far. Similarly,
at the ninth session of UNCTAD, Governments mandated
UNCTAD to continue carrying out its special role in promoting
the integration of trade, environment and development.

51. The Singapore Ministerial Declaration notes that full implementa-
    tion of the WTO Agreements will make an important contribution
    to achieving the objectives of sustainable development. The
    agenda is becoming more balanced and integrated by enlarging
    the development dimension on most issues. Nevertheless, there
    is a perception that mutual understanding between trade,
    environment and development communities is still evolving and
    that a larger consensus still needs to be built on a common
    agenda to strengthen mutual supportiveness of trade, environ-
    ment and development policies. The Singapore Ministerial
    Declaration stressed the importance of policy coordination at the
    national level in the area of trade and environment.

D. CHANGING PRODUCTION AND CONSUMPTION PATTERNS

52. Changing consumption and production patterns in the context of
    sustainable development addresses a broad range of issues,
    including new concepts of economic growth and prosperity,
    efficient use of natural resources, reducing waste, environmen-
    tally sound pricing, product policy and technology transfer.
    Environment and development policy-making aimed at changing
    consumption and production patterns has made noticeable
    progress, in particular in cooperation with international organiza-
    tions and major groups, including business and industry, local
    authorities and the research community. The issue is now promi-

nently placed on the international policy agenda and a number of countries have taken a lead role in facilitating and developing international debate.

53. At the conceptual level, important work has helped to define promising approaches to changing consumption and production patterns, in particular the internalization of environmental costs in goods and services, improved efficiency in energy and materials use and demand-side management. While economic instruments to internalize costs remain difficult to implement, progress has been made in improving efficiency and implementing demand-side management schemes in many industrialized countries. This trend has been promoted by both environmental considerations and the financial benefits of reduced resource and waste flows. Increased attention is being paid by policy makers to social instruments and the provision of adequate infrastructure and facilities, in order to enable individuals to modify their behaviour towards less environmentally damaging patterns. Examples include product labelling, information campaigns and improved recycling schemes.

54. The most promising developments may be seen in the increased participation of non-governmental organizations, business, trade unions, local communities, academics and consumer organizations in efforts to define sustainable levels of consumption and to develop practical programmes of action. However, much work remains to be done in furthering understanding of the possible impacts of changed consumption and production patterns in industrialized countries on the development needs of developing countries.

E. NATURAL RESOURCE MANAGEMENT

1. THE ATMOSPHERE

55. Intensive research has led to scientific consensus within the Intergovernmental Panel on Climate Change (IPCC) that human activities are having a discernible influence on the global climate. The United Nations Framework Convention on Climate Change[8] was one of the key commitments to emerge from UNCED and it has since been ratified by more than 150 States. Many of the parties listed in annex I to the Convention (OECD countries and countries with economies in transition) have developed climate

change action plans involving policy measures, and in some cases targets, for stabilizing or reducing emissions of carbon dioxide and other greenhouse gases. The Montreal Protocol on Substances that Deplete the Ozone Layer, and its subsequent amendments, have already proved effective in reducing emissions of chlorofluorocarbons and have been described as a model for dealing with atmosphere-related issues and for constructive cooperation between Governments, industry, scientists and non-governmental organizations.

56. Despite this progress, CO_2 emissions in most industrialized countries have risen over the past four years and very few countries are likely to stabilize their greenhouse gas (GHG) emissions at 1990 levels by the year 2000. The focus until now has been on technological solutions to increase energy efficiency, which have often been offset by the volume of economic activity. There is still little movement towards strong financial mechanisms that would make fundamental changes in energy consumption possible and no significant new investments have been forthcoming in promoting renewable energy systems.

57. A positive development is the worldwide trend towards increasing competition in the power sector. This will be helpful to small, high efficiency and more economical co-generation systems, while discouraging large, less efficient and less economical stand-alone steam turbine-based power plants. There has also been a noticeable shift in government R&D budgets globally from the fossil energy sector to energy conservation and renewable energy. A further promising initiative is the World Programme on Renewable Energies, launched at the World Solar Summit in Harare in 1996.

58. Transport has become the single largest sectoral end-use of energy in the OECD member countries and is the fastest growing end-use in both developed and developing countries. Transport-related emissions, particularly lead, volatile organic compounds (VOCs) and small particulates now constitute a serious health hazard in many cities worldwide. Initiatives of the Commission on Sustainable Development and international organizations have started the process of phasing out lead in gasoline worldwide. Research continues on alternative vehicle technology, including electric and hybrid vehicles and cleaner fuels, but persistently low fossil fuel prices have discouraged serious development and marketing efforts. Awareness is growing

among authorities and consumers of the financial and health
costs associated with high dependence on motor vehicles and
urban congestion but, to date, there has been little movement
towards strong financial mechanisms and/or economic incentives
to encourage alternative means of transportation. Regulation,
however, is being tightened in most developed countries and
increasingly stringent controls on vehicle emissions are being
introduced, notably in Scandinavia, the European Union and the
United States of America.

2. LAND

59. Land management involves a range of interrelated issues, including land-use planning and employment, habitat preservation, the
maintenance of environmental services such as flood control and
the quality of soils and their fitness for agricultural production.
As competition for land increases, trade-offs between alternative
uses and functions of available land will become more critical in
economic, social and environmental decision-making.

60. The recognition of the need for an integrated approach to
land-use management has increased and was stressed by the
Commission on Sustainable Development at its third session.
Developed countries have made some progress towards integrating agricultural and environmental policies, delinking agricultural
support from production incentives and promoting sustainable
agricultural practices. However, there is still a lack of comprehensive rural policies that bring together production, environmental
and rural welfare objectives. Land resource planning and
management, especially at the implementation phase, are
complex tasks requiring the participation of different
national-level ministries as well as regional and local authorities
and the private sector. More progress is required to develop
institutional arrangements which facilitate joint public–private
activities and improve transparency of land management.

61. There is growing recognition of the need for greater involvement
of all stakeholders concerned in land-use management decisions
and a useful body of experience with participation programmes,
especially in developing countries, is now being built up by
non-governmental organizations, development agencies and
Governments. Many practical programmes relating to programme
design and implementation have been initiated or expanded
since UNCED and a number of countries have made greater

efforts to provide the means for people to express their views on land-use decisions. This process is aided by land resources and development information systems, which have developed rapidly in recent years. Geographical Information Systems are being established in both developed and developing countries, sometimes at the village level.

62. In developing countries, there is a continuing dilemma over production/ income and environmental goals. The strategy of sustainably intensifying already converted land of greatest production potential – in order to reduce pressure for expansion into marginal lands – is beginning to be more widely accepted and introduced. Nevertheless, the importance of non-farm, rural industry promoting policies for employment, especially for areas of lower agricultural potential, are generally not well reflected in rural development and environment strategies.

63. Understanding of the extent and severity of degradation of productive land has been improved by the survey conducted by the International Soil Reference and Information Centre (the GLASOD survey). Following calls for action at UNCED, the United Nations Convention to Combat Desertification in Those Countries Experiencing Serious Drought and/or Desertification, particularly in Africa[9] was opened for signature in October 1994 and entered into force in December 1996. Implementing actions to improve soil management requires a complex of measures involving, according to national circumstances, rationalization of a secure land tenure system, improving farmer education through information and extension programmes, upgrading technology and providing an enabling socio-economic framework which encourages producers to manage their land sustainably.

64. The threat posed to long-term food security by soil degradation was also emphasized at the World Food Summit of November 1996. The Rome Declaration on World Food Security adopted at the Summit states that increased food production must be under-taken within the framework of sustainable management of natural resources and acknowledges the importance for food security of sustainable agricultural practices, fisheries, forestry and rural development. The World Food Summit Plan of Action calls for an ongoing effort to eradicate hunger in all countries, with a minimum target of halving the number of undernourished people by 2015.

65. Land-use conflicts between agriculture, forest cover and urban uses are sharpening, especially in moist tropical areas suitable for the expansion of human activity. The UNCED statement of Forest Principles has helped to encourage global approaches to forest management. The International Tropical Timber Agreement was renegotiated in 1993. A large number of international meetings of experts, many co-sponsored by developing and developed countries, have greatly enriched the understanding of sustainable forest management and of approaches to its implementation. In a major step forward, the Commission on Sustainable Development established the Ad Hoc Intergovernmental Panel on Forests with a two-year mandate to generate consensus and propose actions for implementation of the Forest Principles and other forest-related recommendations of UNCED. The Panel will submit its report to the Commission at its present session. Without pre-empting the outcome of the Panel, it can be said that notable progress has been made towards international consensus on basic principles and operational guidelines for national forest programmes, forest assessment and criteria and indicators for sustainable forest management.

3. Freshwater

66. The recently completed Comprehensive Assessment of the Freshwater Resources of the World has provided new insights into the current status of freshwater availability. It has made clear the intimate relationship between water quantity (supply) and quality and the risks of poor water management. In many developing countries, water scarcity, exacerbated by growing pollution from industry, agriculture and human settlements, constitutes perhaps the most significant threat to socio-economic development and human health.

67. Some progress has been made in developing an integrated approach to water use and more rational and equitable allocation of water among various users. This approach is sometimes characterized by management at the level of river basins or watershed areas and by the participation of users and local communities in the decision-making process, including decisions related to financing of infrastructure. The role of women in water resources management is being increasingly recognized at the national and local levels.

68. Two of the most important success stories in relation to water quality include the development, application and monitoring of drinking water quality guidelines and progress in the eradication of guinea-worm infection. Efforts to improve public water supplies have been ongoing since UNCED and data received from national laboratories indicate that capacities for monitoring water quality are gradually improving. However, water infrastructure in many countries remains wholly inadequate to monitor and control pollution and to protect human health, and current levels of investment do not appear adequate to remedy the situation.

69. A major impediment to the implementation of Agenda 21 objectives remains the fragmentation of responsibilities and mandates for water resources management at the national level and the lack of attention that water receives in comparison with other sectors. The significant economic and social costs associated with poor water quality and inappropriate allocation have yet to be fully realized in decision-making. The lack of financial and human resources also continues to be a major constraint in the improvement of water management capabilities, particularly in developing countries.

70. However, a promising approach is being developed under the Global Water Partnership, an international mechanism which aims to translate the consensus on water management into responsive, coherent services to developing countries, with the emphasis on local implementation. The Partnership will support integrated water resources management programmes by collaborating with Governments and existing networks, developing new arrangements, and encouraging all stakeholders to adopt consistent policies and programmes and to share information and experience.

4. OCEANS AND SEAS

71. Considerable progress has been made in recent intergovernmental negotiations related to oceans and seas. The 1982 United Nations Convention on the Law of the Sea, which entered into force in 1994, and the Agreement for the Implementation of the Provisions of the United Nations Convention on the Law of the Sea relating to the Conservation and Management of Straddling Fish Stocks and Highly Migratory Fish Stocks,[10] which will enter into force after ratification by 30 countries, represent major contributions to the goal of long-term conservation and sustainable use of fish stocks.

72. The adoption, since UNCED, of the Global Programme of Action for the Protection of the Marine Environment from Land-based Activities (Washington, DC, 1995)[11] is another important step towards more integrated management of the world's oceans. The proposals for institutional arrangements for the implementation of the Global Programme of Action are currently being reviewed by Governments. They provide a broad framework for coopera-tion among various United Nations and non-United Nations entities, in particular in the establishment of a clearing-house mechanism and an assessment of the state of oceans and coastal areas. The Global Programme of Action thus complements the Convention on the Prevention of Marine Pollution by Dumping Wastes and Other Matter of 1972 (London Dumping Convention), amended in November 1996.

73. International management of fisheries, strengthened by the entry into force of the United Nations Conference on the Law of the Sea, has improved further with the adoption of General Assembly resolution 46/215, in which the Assembly called for a global moratorium on all large-scale pelagic drift-net fishing on the high seas and with the 1995 agreement reached by the United Nations Conference on Straddling Fish Stocks and Highly Migratory Fish Stocks.[10] These processes, together with the requirements of Agenda 21 for fisheries, were consolidated in the FAO voluntary Code of Conduct for Responsible Fisheries adopted in 1995. Dependence on aquaculture (mainly inland) for an important part of future net increases in fish consumption will require improved management of freshwater resources, protection of aquaculture sites from industrial and urban pollution, as well as protection of coastal areas, wetlands and mangrove swamps from irresponsible coastal aquaculture practices. The need to limit access to marine fishery resources and to establish forms of property and use rights to manage a gradual return to sustainable levels of harvest is gaining recognition, and some countries have begun using individual transferable quotas (ITQs). The International Coral Reef Initiative has addressed the importance of these vulnerable ecosystems and concrete steps are being taken towards implementation.

74. Institutional arrangements for ocean management remain fragmented, however, with problematic divisions of responsibility between areas under national jurisdiction and international waters. It is also evident that, while important agreements have

been concluded at the global level, implementation will be better addressed at the regional level, where the management mandate and capacity of existing organizations needs strengthening. Approximately 80 per cent of marine pollution still stems directly from human activity on land. Protection of the economic and ecological value of coastal ecosystems, as well as of human health, will not be possible without the effective control of pollution from rivers and lakes, and treatment of wastewater from cities which currently drain their urban and industrial wastes directly into coastal systems.

5. Biodiversity

75. The Convention on Biological Diversity,[6] which entered into force in December 1993, has been signed by 163 States and one regional economic integration organization so far. A Subsidiary Body on Scientific, Technical and Technological Advice, as called for in the Convention, has been established. A clearing-house mechanism has been established and is in its pilot phase. It is accessible to all countries and will support implementation of the Convention at the national level. National strategies, plans or programmes for the conservation and sustainable use of biological diversity are under preparation in many countries.

76. At the second meeting of the Conference of the Parties to the Convention, in 1995, UNEP released the Global Biodiversity Assessment which furthered consensus on current trends in biodiversity, means of approaching the problem and possible solutions. In its contribution to the special session of the General Assembly, the Conference of the Parties to the Convention noted that, in spite of the progress made in implementing the objectives of the Convention, Parties remain aware that biological diversity is being destroyed by human activities at unprecedented rates. Despite progress since UNCED, knowledge of biodiversity remains very limited.

77. A Global Strategy for Management of Farm Animal Genetic Resources has been launched with a mission to document existing animal genetic resources, develop and improve their utility to achieve food security, maintain those that represent unique genetic material and that are threatened, and facilitate access to animal genetic resources important to food and agriculture. At its third meeting, in 1996, the Conference of the Parties to the Convention decided to establish a multi-year programme of

activities on agricultural biodiversity aiming, inter alia, at promoting the positive and mitigating the negative effects of agricultural practices on agricultural biodiversity.

78. Work on biosafety has progressed. UNEP issued technical guidelines on biosafety, and under the Convention a Working Group has been established to develop a protocol on biosafety. The International Council of Scientific Unions (ICSU) and the United Nations Educational, Scientific and Cultural Organization (UNESCO) have recently launched an international scientific programme on biodiversity, entitled Diversitas. The Jakarta Mandate, adopted at the second meeting of the Conference of the Parties to the Convention, provides a framework for action on marine and coastal biological diversity. A framework for global action which promotes support for and cooperation with other international bodies was also adopted at the second meeting. At its third meeting, the Conference of the Parties to the Convention elaborated further on actions to be undertaken to advance the implementation of the Convention. It highlighted, among other things, agro-biodiversity, forests and inland water ecosystems.

F. ADDRESSING THE RISKS RELATED TO WASTES AND HAZARDOUS MATERIALS

79. The International Conference on Chemical Safety was organized by the International Programme on Chemical Safety and convened in Stockholm in April 1994. The Conference, which was attended by 110 countries, 10 international organizations and 27 non-governmental organizations, established the Intergovernmental Forum on Chemical Safety, which is mandated to seek consensus among Governments on the development of strategies for the implementation of chapter 19 of Agenda 21 and to undertake periodic reviews of these strategies. The second meeting of the Forum will take place at Ottawa in February 1997.

80. A number of international organizations, namely, FAO, the International Atomic Energy Agency (IAEA), the International Labour Organization (ILO) and the Nuclear Energy Agency (NEA) of OECD, the Pan American Health Organization (PAHO) and the World Health Organization (WHO) have jointly developed and recommended for the use by Governments and industry the International Basic Safety Standards for Protection Against

Ionizing Radiation and for the Safety of Radioactive Sources. Likewise, revised Regulations for the Safe Transportation of Radioactive Material have been set by IAEA.

81. In the field of prior informed consent (PIC), UNEP and FAO are jointly implementing the PIC procedure and negotiations are well under way towards a PIC convention. Two meetings of the Intergovernmental Negotiating Committee were held in 1996 and a diplomatic conference is expected to be held in 1997. The number of countries participating in the voluntary procedure has increased to 148, with 17 chemicals subject to the procedure.

G. ROLE OF GOVERNMENT AND MAJOR GROUPS

82. Agenda 21 makes clear that sustainable development cannot be delivered by Governments alone. It stresses the role of the private sector and other groups in civil society, which have a prominent place in Agenda 21. Experience since 1992 has reinforced the need for such an approach. Globalization affects, and sometimes reduces, the ability of Governments to achieve desired outcomes. While Governments continue to provide the overall framework in which the private sector must operate, many important decisions are made by the private sector, especially by companies operating in an international context. Governments also have to ensure the provision of such basic social services as education and health care, at a time of increasing budget constraints. Practice has shown that detailed prescriptive regulation of the productive sectors is becoming less feasible, less appropriate and less effective. While globalization will have to find a response in new forms of international decision-making, effective implementation of international and national policies requires decentralized and participatory decision-making processes.

1. GOVERNMENTS

83. Close to 150 countries have established national-level commissions or coordinating mechanisms designed to develop an integrated approach to sustainable development and to include a wide range of civil society sectors in the process of agenda setting and strategy building. More than 90 per cent of them have been established in response to UNCED, the majority in developing

countries. In some countries, the national councils of sustainable development have been more political than substantive in nature. They tend to generate broad commitments with limited follow-up at the working level, where sectoral plans and strategies remain largely unaffected.

84. Some of the most promising developments have taken place at the level of cities and municipalities, where local Agenda 21 initiatives have predominated. These have been grass-roots expressions of concern and involvement rather than top-down planning exercises. In many cases local authorities have been reluctant to link their efforts to national action plans for fear that the agenda will then be imposed on them from above rather than flowing from local needs. A wide variety of successful cases have been reported in the past four years on these initiatives. The United Nations Centre for Human Settlements, as part of its preparations for Habitat II, developed, in cooperation with other partners, an extensive database of best practices related to sustainable development at the local level, which is now available on the Internet. Local-level strategies and plans have proved far more successful than those at the national level in terms of making a direct impact.

2. PARLIAMENTS

85. Parliaments in many countries have been actively involved in implementation of the commitments made at UNCED. Information thereon has been reported annually to the Commission on Sustainable Development by the Inter-Parliamentary Union (IPU), based on the outcome of an annual survey. IPU also adopted declarations relevant to sustainable development, for example, on finance and technology transfer and on conservation of world fish stocks.

3. INTERNATIONAL ORGANIZATIONS

86. International cooperation can facilitate the transition towards sustainable development worldwide and support relevant action at the national level. This, together with the growing commitment of intergovernmental organizations and international institutions to the sustainable development agenda, has shown that the United Nations system, in partnership with other international bodies, can, in spite of a number of constraints, add significant value to the implementation of Agenda 21.

87. New forms of cooperation have also emerged at the regional level. They include cooperation between the United Nations regional commissions and representatives of global United Nations agencies and programmes at the regional level. A number of intergovernmental meetings have adopted political statements and action plans for sustainable development. They have been convened in association with both United Nations regional commissions and other regional organizations such as the Organization of American States (OAS), the Organization of African Unity (OAU), the South Pacific Regional Environment Programme (SPREP) and the Association of South-East Asian Nations (ASEAN). A new commission, the Indian Ocean Tuna Commission (IOTC), has been established under FAO for the management of tuna fisheries in the Indian Ocean. High profile ministerial conferences have played an important agenda-setting role and have helped to raise public and political awareness.

88. Regional economic arrangements have rapidly expanded to new countries and new policy areas and have continued to develop after the completion of the Uruguay Round of multilateral trade negotiations. For example, the three States members of the North American Free Trade Agreement (NAFTA) entered into a parallel agreement and established a Commission on Environmental Cooperation to implement it. This framework provides for citizen involvement in monitoring compliance with national environmental laws and regulations.

4. THE PRIVATE SECTOR

89. Agenda 21 has proved to be the starting point for many new business initiatives with sustainability as their stated objective, with notable progress in the areas of joint industry/government partnerships and the development of innovative policy instruments, environmentally efficient technologies and products, and broader sustainability concerns regarding the relation of business and the wider community.

90. Considerations of cost-efficiency and effectiveness are encouraging Governments to supplement traditional regulatory approaches with a broader policy package including economic instruments and other strategies based on "Partnerships" between the private and public sectors. These instruments, which include environmental taxes and charges, environmental subsidies, environmental funds, negotiable emissions instruments,

environmental performance bonds and voluntary agreements, are
in use at some level in all regions of the world by both developed
and developing country Governments.

91. Business and industry organizations are responding with the
development and implementation of an increasing number of
voluntary codes of conduct and environmental management
systems which help firms to meet environmental performance
standards without the need for detailed regulation.
Environmental Management Systems (EMS) standards such as the
ISO 14000 series and the European Union's Environmental
Management and Audit Scheme (EMAS) and other national
standards (such as the United Kingdom BS 7750) have
contributed greatly to industry's adoption of environmental
management and to the ability of Governments to match environmental legislation to industrial improvements in this area. The
International Chamber of Commerce (ICC), UNEP, the United
Nations Industrial Development Organization (UNIDO), UNCTAD
and others are assisting countries, especially developing
countries, in building the capacity necessary to comply with these
new standards, so that they can, inter alia, maintain or enhance
their export opportunities.

92. Many larger companies have now moved beyond "end-of-pipe"
pollution control to a more integrated cleaner production and
life-cycle approach aimed at reducing the environmental impacts
of the goods and services they provide. However, examples on
the market are still limited to a relatively small number of
product categories, notably recycled paper and tissue products,
solvent-free paints and varnishes, cosmetics and more recyclable
packaging. The concept of "eco-efficiency" (the production of
goods and services with reduced throughput of energy and
materials) is being actively promoted by organizations such as the
World Business Council on Sustainable Development and is
attracting growing interest from companies with the resources to
implement technical and managerial change. However, limited
progress has been achieved in addressing small and
medium-sized enterprises. In both developed and developing
countries they are in need of support if they are to manage the
growing sustainability challenge posed by national and international environmental regulations, standards and voluntary codes
that are developed by large companies.

5. OTHER MAJOR GROUPS

93. The action-oriented participation of organized major groups has been particularly dynamic. They increasingly interact directly with national Governments and international organizations, including Convention secretariats and processes. Major group representatives have been active partners in promoting sustainable development among their members and the wider community. Trade unions are bringing sustainable development into the workplace. The scientific and technological communities play a vital role in diagnosing problems and developing response options. Local authorities, by virtue of the geographical focus of their activities, are an increasingly important component of the consultative process regarding local problems and solutions.

94. Youth have been active in promoting sustainable development as vital to their future, though they are still insufficiently included in decision-making at the national and local levels and lack information. Indigenous peoples and farmers are increasingly concerned about the impact of biodiversity and biotechnology issues on traditional values and property rights. Indigenous peoples have become active participants at the intergovernmental level in the context of the Convention on Biological Diversity and the need for protection of genetic resources. Following UNCED, the rights and roles of women in sustainable development have been further emphasized in other global conferences, particularly the Fourth World Conference on Women and the International Conference on Population and Development, and the need to empower women in this regard has been recognized. The rights and role of women are issues that have appeared in all the post-UNCED conferences.

95. Non-governmental organizations are very active in developing and implementing actions for sustainable development at the local and national levels. They increasingly act as implementing partners with national Governments and bilateral and multilateral aid organizations. Non-governmental organizations do not always receive adequate financial support from national institutions and they do not have adequate access to international bodies. The role of non-governmental organizations needs to be enhanced if their full potential in helping to achieve sustainability is to be realized.

96. Despite many positive developments, the implementation of specific objectives in the major group chapters of Agenda 21 has not always achieved the level desired. For example, gender

balance in decision-making has still not been achieved and
national instruments to this effect are not being enforced. The
situation of indigenous people continues to be a serious concern,
with insufficient action being taken at the national level.

H. MEANS OF IMPLEMENTATION

97. A major strength of Agenda 21 lay in its identification of means of
implementation with relevance to different economic sectoral
activities. Approaches to policy implementation currently under
development in these areas are key to building integrative strategies and instruments for sustainable development.

1. FINANCING SUSTAINABLE DEVELOPMENT

98. Average ODA for the period 1993–1995 was lower than for the
period 1990–1992, both in absolute value and as a percentage of
GNP, and was the lowest it had been for 30 years. Only four
countries achieved the goal of 0.7 per cent of GNP. These were
Denmark, the Netherlands, Norway and Sweden. The decrease in
ODA has been particularly critical for the poorest countries that
have little access to other sources of external finance and private
investment. It greatly limits the ability of Governments in most
developing countries to undertake the social and environmental
investments that do not otherwise attract private investments. .
Funds provided under the concessional lending arm of the World
Bank, the International Development Association (IDA), have
been replenished, indicating continued donor commitment to
multilateral cooperation on poverty reduction, economic adjust-
ment and growth and environmental sustainability, though
resources are still not considered large enough. Clearly, the
decline in ODA is not consistent with the expectations raised by
UNCED, despite the efforts to find new and additional sources of
internal finance through alternative mechanisms.

99. In the years since UNCED, the Bretton Woods institutions have
increased their commitment to sustainable development, which
has helped the provision of resources to poor developing
countries for environmentally sound economic and social development. The interest of the World Bank in the environmental
and social impacts of its projects in developing countries has
grown substantially. From marginal concern for environmental

issues before UNCED, the Bank's loan portfolio for environmental projects reached US$ 12 billion in 1996 and it has begun to undertake environmental and social assessments of Bank-financed projects.

100.The Global Environment Facility (GEF) was launched as a pilot programme in 1991 to assist developing countries and countries with economies in transition in pursuit of global benefits in the four focal areas of biodiversity, climate change, international waters, and ozone layer depletion. During the pilot phase, an estimated US$ 730 million was allocated to fund a work programme of 115 global, regional and country projects. In March 1994, an agreement was reached on restructuring and replenishing the Facility as a major source for global environment funding. However, the formulation of proposals and modalities for implementing the projects to be funded by the Facility has often been time-consuming and complicated. Although much work has gone into the preparation of guidelines and proposals, there is room for further improvement in the disbursement of funds in support of GEF projects.

101.The most notable progress in financing sustainable development since UNCED has been the increase in private capital flows to developing countries. The average annual private capital flow to developing countries from OECD countries in the two-year period from 1993 to 1994 was US$ 102 billion or about 60 per cent of total flows from OECD to developing countries. More importantly, about 42 per cent of all private flows from OECD to developing countries in the same period were foreign direct investments, the type of investment that is more stable and reliable in the long term.

102.Despite the increase of private capital flows to middle-level developing countries, the poorest countries have not obtained the necessary flows of private capital and their ratio of foreign direct investment to GNP still remains about half that of the middle-income developing countries.

103.The debt-to-export ratio, the main indicator of an economy's ability to repay its debt, of many developing countries has substantially improved since 1992. The debt problems of the 1980s of most middle-income developing countries have been alleviated through a combination of sound economic policies, liberalization of international trade and capital movements, rescheduling of bilateral external debt and the introduction of new instruments such as Brady-type and debt conversion programmes (of which debt–equity swaps have been the most successful, particularly in Latin America until 1994).

104.However, the debt burden of heavily indebted low-income countries has increased during the past decade, which has hampered their development potential. The initiative taken by the International Monetary Fund (IMF) and the World Bank in April 1996 to design a comprehensive external debt alleviation package targeting these countries is especially welcome.

2. TRANSFERRING TECHNOLOGY

105.Many goals of Agenda 21 depend for their achievement on the introduction of cleaner and more efficient technologies (environmentally sound technologies (ESTs)). The Commission on Sustainable Development, at its third session, adopted a work programme which focuses on access to and dissemination of information, capacity-building for managing technological change and financial and partnership arrangements. Since UNCED, workshops and studies have been conducted and information and awareness-raising campaigns initiated at the national, regional and international levels. They are intended to stimulate the demand for ESTs and thus promote their transfer. A number of developed and developing countries and economies in transition have adopted policies and implemented programmes which support a gradual shift in use from "end-of-pipe" (or clean-up) technologies and equipment to integrated technological solutions in production processes and products. National cleaner production centres that can facilitate the transition towards cleaner production have been established in nine countries with the support of UNIDO and UNEP.

106.Although no concrete data are available, there is overall recognition that the level of technology and technology-related investments from public and private sources in developed countries directed towards developing countries has not, in general, been realized as envisaged at UNCED. Increased private flows have led to investments in industry and technology in some developing countries and economies in transition. However, many developing countries have been left behind, which has slowed the process of technological change in these countries.

107.More information is needed from both national and local governments and the private sector regarding the effectiveness of

policies to facilitate and accelerate technology transfer and
technological diffusion. Such information could provide greater
insights into (a) the relationship between environmental
concerns and the demand for technologies and technical innova-
tion; (b) the effectiveness of company strategies for adapting to
the requirements of technological change and support for
production processes which are environmentally responsible and
competitive; and (c) trends regarding the dynamics of national
environmental technology markets and more accurate international data regarding technology flows to developing countries.

3. BUILDING CAPACITY

108. The emphasis in Agenda 21 on more participatory approaches to
sustainable development has influenced a new generation of
capacity-building projects that has come on line since 1992. Most
activities aimed at environmental management and sustainable
development now make explicit efforts at stakeholder and benefi-
ciary assessment. For example, the Capacity 21 programme
established by UNDP after UNCED has proved to be an effective
catalyst and learning mechanism to support capacity-building for
sustainable development. As of May 1996, total contributions to
Capacity 21, through both its trust fund and other mechanisms,
stood at around US$ 57 million. Since 1993, the programme has
helped to fund projects in over 40 countries.

109. Much progress has been made in the areas of strategy formulation, greater participation and information exchange. What has
been lacking are the structures and capacities to carry out many
of the technical functions associated with sustainable development. Putting such technical, scientific and institutional
structures in place represents the long-term work facing many
countries. The lack of co-financing from bilateral donors has
slowed their ability to work on the larger multi-component
capacity-building programmes.

4. INFORMATION FOR DECISION-MAKING

110. Good quality information and data are crucial in identifying the
nature and scale of problems but progress in collecting, organizing and presenting information in usable form has been mixed.
The quality of information at the international level, in terms of
data collection and development of indicators, has improved
considerably since UNCED. National and local level information,

and facilities for the exchange of such information, need to be further developed and improved.

111. Many of the data areas identified in Agenda 21, including urban air, freshwater, desertification, biodiversity, high seas and upper atmosphere, demographic factors, urbanization, poverty, health, rights of access to resources, and information on various major groups, have been inventoried at the international and regional levels. Considerable progress has been made in filling gaps, through initiatives of the United Nations system, other intergovernmental organizations and non-governmental organizations. To address the lack of critical long-term data necessary to understand global ecosystem problems, international organizations and the scientific community have designed observing systems to make data collection more coherent and cost-effective. Important mechanisms for observation, monitoring, assessment and exchange have been put in place to assess the state of planetary systems and enhance information flow. These include the Global Terrestrial Observing System (G-TOS), the Global Climate Observing System (G-COS) and the Global Ocean Observing System (GOOS). Notable innovations since UNCED include the World Hydrological Cycle Observing System (WHYCOS), the Global Coral Reef Monitoring Network, the Mountain Forum and the Global Modelling Forum. Important work is being carried out, in particular under the auspices of the Intergovernmental Forum on Chemical Safety (IFCS), on the development of information on chemical safety.

112. Since UNCED, much new work has been initiated on indicators for sustainable development. The Commission on Sustainable Development, for its part, launched a global process to draw upon these initiatives and make use of their collective expertise and knowledge to reach consensus on the technical validity, comparability and political acceptability of indicators. The Commission approved a programme of work which has led to the development of a preliminary core set of indicators of sustainable development, followed by the preparation of methodology sheets for each of the indicators. The aim is to have an agreed set of indicators available for use at the national level by the year 2000.

113. At the same time, work is progressing in various sectors to develop more detailed sectoral indicators to measure performance under international agreements, and in the scientific community to integrate the economic, social, environmental and

institutional dimensions in more aggregated indicators, which take account of interlinkages. In this regard, a particularly fruitful approach to integrating the components of sustainable development in an operational framework is to think in terms of flows of investment which maintain or increase society's stock of environmental assets (natural capital), physical capital (the built environment), human capital and social capital. While limited substitution is possible among these different categories of assets, they are for the most part complementary. The challenge of sustainable development is thus to build up all kinds of wealth into a people-enriching and nature-preserving system.

114.Less, though significant, progress has been made at the national and subnational levels. A growing number of countries have completed national inventories and organized the collection of needed data. Several factors account for this trend, including the rapid growth of national and subnational sustainable development strategies, plans and targets; adoption of national and local indicators; ratification of relevant international treaties; and, in some cases, support from the international community for the requisite capacity-building for these activities.

115.The work that has begun on streamlining national reporting in the field of sustainable development is of considerable importance and should be continued. Emphasis should be placed both on establishing a multi-year work programme that focuses, inter alia, on a calendar of reporting to assist national planning and on information-sharing among United Nations system organizations through electronic means to the extent possible.

116.Overall, great strides have been made in the availability of information independently of Agenda 21, as a result of rapid and revolutionary technological changes in computing, telecommunications and geographical information system technologies. However, far too little has been done to make national telecommunications systems responsive to the growing demand for electronic information. This is especially the case in some developing countries, where the lack of adequate infrastructure and telephone systems is hindering access to the new electronic networks.

III. CHALLENGES AND PRIORITIES AHEAD

117. The previous sections lead towards a number of general conclusions that should be taken into account when defining priorities for future international action towards sustainable development. Progress is evident in the many plans and strategies which have been developed at every level of operation. Strategies are the first step in the policy cycle which must then advance to politically difficult decisions on priorities and budget allocation, implementing actions and review.

118. Much remains to be done to ensure that sustainable development is understood by decision makers as well as by the public. Accordingly, there is a need for adequate communication strategies at the international and country levels to ensure that this understanding is achieved.

119. The four years of implementation of Agenda 21 have underlined the crucial importance of an integrated approach to sustainable development, involving all actors in a participatory process. Sustainable development strategies are important mechanisms to enhance and link national capacity, bringing together the priorities in social, economic and environmental policies involving participation of all concerned parties. Capacity-building activities should give priority to their development and implementation. Such strategies should also extend to the various levels of government. Effective planning and implementation of sustainable development policies requires the participation of all social groups. Responsibility for managing resources, particularly at the local level, are often divided between women and men and between different socio-economic groups. Each group has specific knowledge and skills that should be integrated in the planning process, and participation in the policy-making process will encourage broad commitment to, and implementation of, sustainable development policies.

120. Eradication of poverty throughout the world is a priority element of sustainable development, as stated in Agenda 21 and further elaborated in the Programme of Action of the World Summit for Social Development.[12] For people living in poverty, and for countries with a high incidence of poverty, the eradication of poverty must be a high priority, both as an end in itself and to promote sustainable use of natural resources. The commitments of the international community to support the efforts of developing countries must be reaffirmed and implemented.

121. Substantive progress has been achieved in further developing and adopting international consensus on the sustainable management of natural resources in the form of international agreements. The focus of international discussions, including those taking place under the Rio Conventions (on climate change,[8] biodiversity[6] and desertification[9]), has moved from policy development to implementation. This holds true for the atmosphere, oceans, land management and biodiversity.

122. However, further policy development at the global level as guidance for implementation is needed in some areas. The Ad Hoc Intergovernmental Panel on Forests is likely to leave certain areas of sustainable forest management unresolved. Evidence provided by the Comprehensive Assessment of the Freshwater Resources of the World sheds new light on the urgency of the freshwater situation in the world, which requires a consolidated policy response. Scientific evidence on the negative impacts on health and ecosystems of certain chemicals, especially POPs, is such that the need for an international agreement on their phase-out is urgent.

123. In certain areas there is a need to improve policy coordination and implementation at the regional level. Some issues, such as regional seas, certain aspects of climate change, transboundary conservation of biodiversity, transboundary environmental impacts, land degradation and transboundary movements of hazardous wastes, are among those which can be best tackled at the regional level.

124. The need for an integrated approach in the management of each of the natural resources is one of the important lessons from UNCED and its follow-up. In some areas, for example, oceans, a number of agreements have come into being that are not necessarily interlinked. Follow-up discussions must ensure improved integration.

125. More can be done to make the implementation of the three Rio Conventions (on climate change, biodiversity and desertification) mutually reinforcing, by addressing substantive linkages and identifying projects that achieve the objectives of more than one Convention.

126. To facilitate effective implementation, consideration of resource management issues must be combined with an equal emphasis on sectoral policy development. Economic sectors (agriculture, fisheries, forestry, industry, human settlements, energy, transport,

social services) must be involved in international discussions on implementation and held responsible for their contribution to problems and solutions. In this context, due attention should be given to such issues as health, sound management of wastes and chemical safety, among others.

127. Urgent action is required to slow, and where feasible, reverse the degradation of agricultural land. Improved management and restoration of irrigated land, and improved land-use planning to reduce unnecessary losses of productive land to development, are priorities. Future food supplies will come in large part from intensification of agriculture, that is, increased yields from existing lands. Soil degradation and loss of productive lands reduce the potential for future gains and increase the technological, social and financial challenges of raising production.

128. The integration of health impact assessments into economic sectoral planning and in sustainable development plans needs priority attention.

129. Major gaps in international discussion of economic sectors exist, namely in the fields of energy, transport and tourism. Energy is arguably the most critical link between environment and development, but the tensions between the legitimate energy needs of developing countries for socio-economic development and the consequences of expanded use of fossil fuels for human health and local, regional and global pollution have been inadequately addressed. This oversight has to some extent been redressed in the negotiations on climate change, but it remains an area that requires more focused analysis and action, not just in terms of new and renewable forms of energy, but the more basic issue of how developing countries, in particular, can acquire the levels of energy supply needed for their development while reducing dependence on carbon-based fuels. In transport, improvements in efficiency in fuel and material use are largely outweighed by the growth of the sector. Tourism is the fastest growing economic sector, with major social and environmental impacts.

130. The discussion on changing consumption and production patterns must move from a rather abstract level, for example, by providing a strategic approach and concrete measures to be taken by the economic sectors. The impact of changing consumption and production patterns in industrialized countries on the export opportunities of developing countries must be kept under permanent review.

131. The need for international discussions on making the objectives of trade liberalization and sustainable development mutually supportive is undisputed. This should be coupled with enhanced discussion and coordination at the national level. The focus of the debate could shift from narrowly defined trade and environment issues to an integrated consideration of all factors relevant for achieving sustainable development, with an emphasis on synergies rather than on restrictions. The debate should be supported by improved empirical analysis. Identification and effective implementation of positive measures deserve priority attention.

132. Broad participation has been essential for achieving progress in policy development and implementation of sustainable development. Further consideration must be given to new forms of governance that reflect the increased responsibility and accountability of major groups. The role of the private sector is ever increasing. Major resource flows from the developed to the developing countries and economies in transition take place through the private sector. Although private capital has the potential to finance sustainable development, so far it has typically avoided projects whose main purpose is to generate environmental and social benefits. However, developing countries offer investment opportunities that generate social and environmental gains and could also be profitable if they ensure a more efficient provision of goods and services for which users are willing to pay (win-win opportunities). Furthermore, involvement of the private sector (industry and the private financial sector) in the international policy discussions on sustainable development is needed.

133. Public international financial resources for the implementation of sustainable development in developing countries have not met the commitments made by the donor countries at UNCED. Reconfirmation of the commitments made at UNCED and specific commitments for support to those areas that are closely linked to the fulfilment of basic needs and where coordinated programmes have been developed by the international community (e.g., in the field of water, energy and forests) are needed to maintain the credibility of partnerships between developed and developing countries. The scarcity of funds to finance sustainable development is particularly acute for low-income developing countries. This is because they attract little external private capital, receive

decreasing amounts of ODA, and many of them have heavy external debt burdens. The 1990s have witnessed growing gaps between least developed countries and other developing countries with regard to GNP and per capita income growth rates, and many other indicators of human development. It would appear that developing countries that have embraced sound, stable and outward-oriented macroeconomic and trade policies during the 1990s (such as those adopted by many middle-income Latin American and Asian countries) do attract private capital and have easier access to external debt alleviation programmes. However, these policies require costly political, economic and administrative reforms. Because ODA is an important source of finance for these reforms, especially in the least developed countries, donor countries should intensify efforts to meet the UNCED target on ODA.

134.Limited progress has been made with the implementation of economic instruments to internalize environmental costs in goods and services. Active exchange of information on the successful use of economic instruments may be conducive to their further introduction.

135.Technology partnerships and cooperative arrangements are needed to stimulate practical cooperation between Governments and industry at both the national and international levels. More information is needed from both Governments and the private sector regarding the effectiveness of policies to facilitate and accelerate technology transfer and technological diffusion.

136.The information base for decision-making on sustainable development is still uneven and access to existing information systems must be strengthened. There is a need to improve capabilities for data acquisition from large areas of the world on a scale necessary to monitor the environment properly. Many of the environmental problems, such as climate change, desertification and the extinction of living species, unfold over long timescales. The need for data and processed information with adequate temporal and spatial resolution is still significant.

137.Consolidated scientific evidence is essential for international policy development. There is a need for further scientific cooperation, especially across academic disciplines, in order to verify and strengthen scientific evidence for environmental change. Some examples of cooperation already exist, such as the Intergovernmental Panel on Climate Change, and work carried out by the Intergovernmental Forum on Chemical Safety. Education for all needs to be assured as another crucial factor associated with policy development. The concerns of sustainable development, global interdependence and peace must be fully integrated in formal and non-formal education and public awareness-raising.

138.A gender perspective should be applied in all aspects of the implementation of Agenda 21. This is essential in order to assess the actual and potential contribution of women and men to formulating and implementing relevant policies and programmes, as well as to adequately assess the impacts of economic and social conditions and of environmental degradation on the population as a whole. The need for gender-disaggregated data is a priority in facilitating gender-sensitive analysis and policy-making.

IV. INSTITUTIONAL FRAMEWORK AND THE ROLE OF THE COMMISSION ON SUSTAINABLE DEVELOPMENT AFTER 1997

A. INSTITUTIONAL FRAMEWORK

139.The collective view of the secretariats of the organizations of the United Nations system is that the concept of sustainable development should continue to provide an "overarching" policy framework for the entire spectrum of United Nations activities in the economic, social and environmental fields at the global, regional and national levels. All intergovernmental and inter-agency bodies and processes should contribute, within their mandates and areas of competence, to further progress in achieving the goals of sustainable development through concrete action and decision-making. Full account must also be taken of the overall framework for cooperation agreed by the international community in the context of the coordinated follow-up to all recent United Nations conferences, since all have made an important contribution to specific aspects of the global sustainable development agenda.

140.The overall institutional framework for the implementation of Agenda 21, as outlined in chapter 38, would seem to be fully relevant for the period after the 1997 review. However, the

General Assembly at its special session may wish to consider how this framework could best be deployed in the future. More specific suggestions on this matter are contained in document E/CN.17/1997/2/Add.28. At the same time, bearing in mind the specific request contained in paragraph 13 (d) of General Assembly resolution 50/113, the present report includes recommendations on the future role of the Commission on Sustainable Development in the follow-up to the special session, which are outlined below.

B. Programme of work of the Commission on Sustainable Development

141. The first multi-year programme of work of the Commission on Sustainable Development was organized in a way that allowed for in-depth consideration of all the individual chapters of Agenda 21 over the period of three years. Such an approach was appropriate for the first review cycle. It provided an effective opportunity for the Commission to carry out an initial analysis of institutional and policy changes and activities at the international, national and "major group" levels to implement all chapters of Agenda 21, adopt specific recommendation to operationalize specific recommendations of UNCED and provide a forum for exchanges of relevant experiences.

142. However, some disadvantages have become apparent. Annual sessions of the Commission have been overloaded with issues and reports. Additionally, separate consideration of individual chapters of Agenda 21 has not always allowed the Commission to examine linkages between various sectoral and cross-sectoral issues addressed in Agenda 21, and interrelationships among various economic, social and environmental aspects of sustainable development have not always been addressed adequately. The impression has sometimes been created that there is some duplication of work between the Commission and other intergovernmental bodies or processes. At the same time, the annual, generic policy discussion of some issues (e.g., the role of major groups; economic instruments; decision-making, capacity-building) has become somewhat repetitive.

143. The experience of the first programme of work of the Commission should be taken into account in designing the next

cycle of the work programme. In addition, the substantive results
of the assessment of overall progress in implementing Agenda 21,
and future priorities to be identified at the special session, would
need to be considered.

144.Following the 1997 review, the Commission should continue to
provide a central forum for reviewing further progress in the
implementation of Agenda 21 and for policy debate on sustainable development in general. At the same time, it would seem
essential for the Commission to ensure a greater focus on those
issues that require further policy discussion and agreement.
Comprehensive reviews of all chapters of Agenda 21 could be
carried out only once in several years, or as the need arises.

145.It is suggested that the future programme of work of the
Commission could be organized on the basis of the following
considerations:

(a) Implementation of all chapters/thematic areas of Agenda 21
will continue to be reviewed. However, only a limited number
of chapters/thematic areas will be given in-depth consideration in a given year. Other chapters/thematic areas will be
reviewed only in the context of their relationship to those
chapters/areas which are the current focus of discussion. In
other words, chapters/thematic areas chosen in a given year
for a focused, in-depth discussion could serve as an "entry
point" to a broader discussion, involving linkages with related
provisions under other chapters (both in terms of conceptual
and/or policy linkages, and in terms of consideration of
relevant means of implementation);

(b) The following criteria may be applied for selecting those issues
which would be the subject of focused discussion in the next
multi-year programme of work of the Commission:

(i) Issues should be of significance in achieving the goals of
sustainable development worldwide, involving promotion
of policies which integrate economic, social and environmental dimensions of sustainability and promote
coherence of action at all levels;

(ii) Issues should require further dialogue and
consensus-building before internationally agreed strategies or frameworks for action could be adopted;

(iii) Issues should be "cross-cutting", providing the opportunity for integrated consideration;

(iv) Issues should involve means of implementation, the role of various economic sectors and major groups, and matters relating to socio-economic factors, such as health or consumption and production patterns, which need to be given a more prominent role in the work programme. However, consideration of these issues would be better integrated into the discussion of specific thematic areas;

(v) Issues which are addressed in Agenda 21, but which are dealt with in a systematic way in another intergovernmental body/process (i.e., human settlements (Commission on Human Settlements); poverty (Commission for Social Development); climate change (United Nations Framework Convention on Climate Change); biodiversity (Convention on Biological Diversity); desertification (United Nations Convention to Combat Desertification in Those Countries Experiencing Serious Drought and/or Desertification, Particularly in Africa)) may not be the subject of separate focused discussions in the Commission, but should be considered only in terms of their relationship to other issues;

(c) To lighten the Commission's agenda and to provide for a more focused and in-depth consideration of key policy issues, the Commission, during its next programme cycle, could limit itself to only three substantive items on the agenda of its annual sessions. This would allow the Commission, during the next four years (1998–2001) to consider thoroughly all issues that will be selected for in-depth discussion and, at the same time, undertake an integrated analysis of all of the chapters of Agenda 21. In 2002, the Commission could carry out a second comprehensive review of overall progress in the implementation of Agenda 21 in its entirety.

146. More specifically, in a given year the Commission could include three substantive items on its agenda[1]:

(a) An item dealing with a cluster of cross-sectoral issues;

(b) An item focusing on sustainable development in a natural resource sector;

(c) An item focusing on the role of a relevant economic sector/major group in sustainable development.

147. The suggested timing for the consideration of various "key" themes takes into account the time when similar issues were considered in the Commission during the period 1993–1996, relevant decisions taken by the Commission and other intergov-

ernmental bodies that deal with the consideration of specific issues after 1997, and expected outcomes of ongoing intergovernmental processes. The General Assembly at its special session may wish to consider a different schedule. Special consideration was also given to conceptual linkages between various issues, inter alia, with a view to attracting attention to the work of the Commission of ministers and national policy makers responsible for specific economic sectors, who may wish to attend the high-level segments of the Commission in a given year jointly with ministers of environment and development.

NOTES

1   Report of the United Nations Conference on Environment and Development, Rio de Janeiro, 3–14 June 1992, vol. I, Resolutions Adopted by the Conference (United Nations publication, Sales No. E.93.I.8 and corrigendum), resolution 1, annex II.
2   Ibid., annex I.
3   Ibid., annex III.
4   World Population Prospects: 1996 Revision (United Nations publication, forthcoming).
5   World Population Prospects: 1994 Revision (United Nations publication, Sales No. E.95.XIII.16).
6   See United Nations Environment Programme, Convention on Biological Diversity (Environmental Law and Institutions Programme Activity Centre), June 1992.
7   Report of the Global Conference on the Sustainable Development of Small Island Developing States, Bridgetown, Barbados, 25 April–6 May 1994 (United Nations publication, Sales No. E.94.I.18 and corrigenda), chap. I, resolution 1, annex II.
8   A/AC.237/18 (Part II)/Add.1, annex I.
9   A/49/84/Add.2, annex, appendix II.
10 A/50/550, annex I.
11 A/51/116, annex I, appendix II.
12 Report of the World Summit for Social Development, Copenhagen, 6–12 March 1995 (United Nations publication, Sales No. E.96.IV.8), chap. I, resolution 1, annex II.

Annex 4

PROGRAMME FOR THE FURTHER IMPLEMENTATION OF AGENDA 21

ADOPTED BY THE SPECIAL SESSION OF THE GENERAL ASSEMBLY 23–27 JUNE 1997

CONTENTS

A. STATEMENT OF COMMITMENT

1. At the nineteenth special session of the United Nations General Assembly, we – Heads of States and Governments and other Heads of Delegations, together with our partners from international institutions and non-governmental organizations, have gathered to review progress achieved over the five years that have passed since the United Nations Conference on Environment and Development and to re-energize our commitment to further action on goals and objectives set out by the Rio Earth Summit.

2. The United Nations Conference on Environment and Development was a landmark event. At that Conference, we launched a new global partnership for sustainable development – a partnership which respects the indivisibility of environmental protection and the development process. It is founded on a global consensus and political commitment at the highest level. Agenda 21, adopted at Rio, addresses the pressing environment and development problems of today and also aims at preparing the world for the challenges of the next century to attain the long-term goals of sustainable development.

3. Our focus at this special session has been to accelerate the implementation of Agenda 21 in a comprehensive manner and not to re-negotiate its provisions or to be selective in its implementation. We reaffirm that Agenda 21 remains the fundamental programme of action for achieving sustainable development. We reaffirm all the principles contained in the Rio Declaration on Environment and Development and the Forest Principles. We are convinced that the achievement of sustainable development requires the integration of its economic, environmental and social components. We re- commit to work together – in the spirit of global partnership – to reinforce our joint efforts to meet equitably the needs of present and future generations.

4. We acknowledge that a number of positive results have been achieved, but we are deeply concerned that the overall trends for sustainable development are worse today than they were in 1992. We emphasize that the implementation of Agenda 21 in a comprehensive manner remains vitally important and is more urgent now than ever.

5. Time is of the essence to meet the challenges of sustainable development as set out in the Rio Declaration and Agenda 21. To this end, we re-commit ourselves to the global partnership estab-

lished at UNCED and to the continuous dialogue and action inspired by the need to achieve a more efficient and equitable world economy, as a means to provide a supportive international climate for achieving environment and development goals. We, therefore, pledge to continue to work together in good faith and in the spirit of partnership, to accelerate the implementation of Agenda 21. We invite everyone throughout the world to join us in our common cause.

6. We commit ourselves to ensure that the next comprehensive review of Agenda 21 in the year 2002 demonstrates greater measurable progress in achieving sustainable development. This programme for the further implementation of Agenda 21 is our vehicle to achieve that. We commit ourselves to fully implement this programme.

B. Assessment of progress made since the United Nations Conference on Environment and Development

7. The five years that have elapsed since the United Nations Conference on Environment and Development (UNCED)[1] have been characterized by the accelerated globalization of interactions among countries in the areas of world trade, foreign direct investment and capital markets. Globalization presents new opportunities and challenges. It is important that national and international environmental and social policies be implemented and strengthened in order to ensure that globalization trends have a positive impact on sustainable development, especially in developing countries. The impact of recent trends in globalization on developing countries has been uneven. A limited number of developing countries have been able to take advantage of those trends, attracting large inflows of external private capital and experiencing significant export-led growth and acceleration of growth in per capita gross domestic product (GDP). Many other countries, however, in particular African countries and the least developed countries, have shown slow or negative growth and continue to be marginalized. As a result, they generally experienced stagnating or falling per capita GDP through 1995. In these and in some other developing countries, the problems of poverty, low levels of social development, inadequate infra-structure and lack of capital have prevented them from benefiting

from globalization. While continuing their efforts to achieve sustainable development and to attract new investments, these countries still require international assistance in their efforts towards sustainable development. In particular the least developed countries continue to be heavily dependent on a declining volume of official development assistance (ODA) for the capacity-building and infrastructure development required to provide for basic needs and more effective participation in the globalizing world economy. In an increasingly interdependent world economy, the responsible conduct of monetary and other macroeconomic policies requires that their potential impact on other countries be taken into account. Since UNCED, the countries with economies in transition have achieved significant progress in implementing the principles of sustainable development. However, the need for full integration of these countries into the world economy remains one of the crucial problems on their way towards sustainable development. The international community should continue to support these countries in their efforts to accelerate the transition to a market economy and to achieve sustainable development.

8. Although economic growth – reinforced by globalization – has allowed some countries to reduce the proportion of people in poverty, marginalization has increased for others. Too many countries have seen economic conditions worsen and public services deteriorate; the total number of people in the world living in poverty has increased. Income inequality has increased among countries and also within them, unemployment has worsened in many countries, and the gap between the least developed countries and other countries has grown rapidly in recent years. On a more positive note, population growth rates have been declining globally, largely as a result of expanded basic education and health care. That trend is projected to lead to a stable world population in the middle of the twenty-first century. There has also been progress in social services, with expanding access to education, declining infant mortality and increasing life expectancy in most countries. However, many people, particularly in the least developed countries, still do not have access to adequate food and basic social services or to clean water and sanitation. Reducing current inequities in the distribution of wealth and access to resources, both within and among countries, is one of the most serious challenges facing humankind.

9. Five years after UNCED, the state of the global environment has continued to deteriorate, as noted in the Global Environment Outlook of the United Nations Environment Programme (UNEP),[1] and significant environmental problems remain deeply embedded in the socio-economic fabric of countries in all regions. Some progress has been made in terms of institutional development, international consensus-building, public participation and private sector actions and, as a result, a number of countries have succeeded in curbing pollution and slowing the rate of resource degradation. Overall, however, trends are worsening. Many polluting emissions, notably of toxic substances, greenhouse gases and waste volumes are continuing to increase although in some industrialized countries emissions are decreasing. Marginal progress has been made in addressing unsustainable production and consumption patterns. Insufficient progress has also been identified in the field of environmentally sound management and adequate control of adequate transboundary movements of hazardous and radioactive wastes. Many countries undergoing rapid economic growth and urbanization are also experiencing increasing levels of air and water pollution, with accumulating impacts on human health. Acid rain and transboundary air pollution, once considered a problem only in the industrialized countries, are increasingly becoming a problem in many developing regions. In many poorer regions of the world, persistent poverty is contributing to accelerated degradation of natural resources and desertification has spread. In countries seriously affected by drought and or desertification, especially those in Africa, their agricultural productivity, among other things, is uncertain and continues to decline, thereby hampering their efforts to achieve sustainable development. Inadequate and unsafe water supplies are affecting an increasing number of people worldwide, aggravating problems of ill health and food insecurity among the poor. Conditions in natural habitats and fragile ecosystems, including mountain ecosystems, are still deteriorating in all regions of the world, resulting in diminishing biological diversity. At the global level, renewable resources, in particular freshwater, forests, topsoil and marine fish stocks, continue to be used at rates beyond their viable rates of regeneration; without improved management, this situation is clearly unsustainable.

10. While there has been progress in material and energy efficiency, particularly with reference to non-renewable resources, overall trends remain unsustainable. As a result, increasing levels of pollution threaten to exceed the capacity of the global environment to absorb them, increasing the potential obstacles to economic and social development in developing countries.

11. Since UNCED, extensive efforts have been made by Governments and international organizations to integrate environmental, economic and social objectives into decision-making by elaborating new policies and strategies for sustainable development or by adapting existing policies and plans. As many as 150 countries have responded to the commitments established at UNCED through national-level commissions or coordinating mechanisms designed to develop an integrated approach to sustainable development.

12. The major groups have demonstrated what can be achieved by taking committed action, sharing resources and building consensus, reflecting grass- roots concern and involvement. The efforts of local authorities are making Agenda 21 and the pursuit of sustainable development a reality at the local level through the implementation of "Local Agenda 21s" and other sustainable development programmes. Non-governmental organizations, educational institutions, the scientific community and the media have increased public awareness and discussion of the relations between environment and development in all countries. The involvement, role and responsibilities of business and industry, including transnational corporations, are important. Hundreds of small and large businesses have made "green business" a new operating mode. Workers and trade unions have established partnerships with employers and communities to encourage sustainable development in the workplace. Farmer-led initiatives have resulted in improved agricultural practices contributing to sound resource management. Indigenous people have played an increasing role in addressing issues affecting their interests and particularly concerning their traditional knowledge and practices. Young people and women around the world have played a prominent role in galvanizing communities to recognize their responsibilities to future generations. Nevertheless, more opportunities should be created for women to participate effectively in economic, social and political development as equal partners in all sectors of the economy.

13. Among the achievements since UNCED are the entry into force of the United Nations Framework Convention on Climate Change (A/AC.237/18 (Part II)/Add.1 and Corr.1, annex I), the Convention on Biological Diversity[2] and the United Nations Convention to Combat Desertification in Those Countries Experiencing Serious Drought and/or Desertification, Particularly in Africa (A/49/84/Add.2, annex, appendix II); the conclusion of an agreement on straddling and migratory fish stocks (A/50/550, annex I); the adoption of the Programme of Action for the Sustainable Development of Small Island Developing States;[3] and the elaboration of the Global Programme of Action for the Protection of the Marine Environment from Land-based Activities (A/51/116, annex II) and the entry into force of the United Nations Convention on the Law of the Sea (UNCLOS).[4] Implementation of these important commitments and of others adopted before UNCED by all the parties to them, however, remains to be carried out, and in many cases further strengthening of their provisions is required as well as the mechanisms for putting them into effect. The establishment, restructuring, funding and replenishment of the Global Environment Facility (GEF) were a major achievement. However, its levels of funding and replenishment have not been sufficient fully to meet its objectives.

14. Progress has been made in incorporating the principles contained in the Rio Declaration on Environment and Development[5] – including the principle of common but differentiated responsibilities, which embodies the important concept of and basis for international partnership; the precautionary principle; the polluter pays principle; and the environmental impact assessment principle – in a variety of international and national legal instruments. While some progress has been made in implementing UNCED commitments through a variety of international legal instruments, much remains to be done to embody the Rio Principles more firmly in law and practice.

15. A number of major United Nations conferences have advanced international commitment for the achievement of long-term goals and objectives towards sustainable development.

16. Organizations and programmes of the United Nations system have played an important role in making progress in the implementation of Agenda 21. The Commission on Sustainable Development was established to review progress achieved in the implementation of Agenda 21, advance global dialogue and foster

partnerships for sustainable development. The Commission has catalyzed new action and commitments and has contributed to the deliberations on sustainable development among a wide variety of partners within and outside the United Nations system. Although much remains to be done, progress has also been made at the national, regional and international levels in implementing the UNCED Forest Principles,[6] including through the Commission's Ad Hoc Intergovernmental Panel on Forests.

17. Provision of adequate and predictable financial resources and the transfer of environmentally sound technologies to developing countries are critical elements for the implementation of Agenda 21. However, while some progress has been made, much remains to be done to activate the means of implementation set out in Agenda 21, in particular in the areas of finance and technology transfer, technical assistance and capacity-building.

18. Most developed countries have still not reached the United Nations target, reaffirmed by most countries at UNCED, of committing 0.7 per cent of their gross national product (GNP) to official development assistance or the United Nations target, as agreed, of committing 0.15 per cent of GNP as ODA to the least developed countries. Regrettably, on average, ODA as a percentage of the GNP of developed countries has drastically declined in the post-UNCED period, from 0.34 per cent in 1992 to 0.27 per cent in 1995, but ODA has taken more account of the need for an integrated approach to sustainable development.

19. In other areas, results have been encouraging since UNCED. There has been a sizeable expansion of private flows of financial resources from developed to a limited number of developing countries and, in a number of countries, efforts have been made in support of domestic resource mobilization, including the increasing use of economic instruments to promote sustainable development.

20. In many developing countries, the debt situation remains a major constraint to achieving sustainable development. Although the debt situation of some middle-income countries has improved, there is a need to continue to address the debt problems of the heavily indebted poor countries, which continue to face unsustainable external debt burdens. The recent World Bank/International Monetary Fund (IMF) Heavily Indebted Poor Countries Initiative could help to address that issue with the

cooperation of all creditor countries. Further efforts by the international community are still required to remove debt as an impediment to sustainable development.

21. Similarly, technology transfer and technology-related investment from public and private sources, which are particularly important to developing countries, has not been realized as outlined in Agenda 21. Although increased private flows have led to investments in industry and technology in some developing countries and economies in transition, many other countries have been left behind. Conditions in some of these countries have been less attractive to private sector investment and technological change has been slower, thus limiting their ability to meet their commitments to Agenda 21 and other international agreements. The technology gap between developed countries and, in particular, the least developed countries has widened.

C. IMPLEMENTATION IN AREAS REQUIRING URGENT ACTION

22. Agenda 21 and the principles contained in the Rio Declaration on Environment and Development established a comprehensive approach to the achievement of sustainable development. While it is the primary responsibility of national Governments to achieve the economic, social and environmental objectives of Agenda 21, it is essential that international cooperation be reactivated and intensified, recognizing, inter alia, the principle of common but differentiated responsibilities as stated in principle 7 of the Rio Declaration. This requires the mobilization of stronger political will and the invigoration of a genuine new global partnership, taking into account the special needs and priorities of developing countries. That approach remains as relevant and as urgently needed as ever. It is clear from the assessment above that, although progress has been made in some areas, a major new effort will be required to achieve the goals established at UNCED, particularly in areas of cross-sectoral matters where implementation has yet to be achieved. The proposals set out in sections 1–3 below outline strategies for accelerating progress towards sustainable development. The sections are equally important and must be considered and implemented in a balanced and integrated way.

1. INTEGRATION OF ECONOMIC, SOCIAL AND ENVIRONMENTAL OBJECTIVES

23. Economic development, social development and environmental protection are interdependent and mutually reinforcing components of sustainable development. Sustained economic growth is essential to the economic and social development of all countries, in particular developing countries. Through such growth, which should be broadly based so as to benefit all people, countries will be able to improve the standards of living of their people through the eradication of poverty, hunger, disease and illiteracy and the provision of adequate shelter and secure employment for all, and the preservation of the integrity of the environment. Growth can foster development only if its benefits are fully shared. It must therefore also be guided by equity, justice and social and environmental considerations. Development, in turn, must involve measures that improve the human condition and the quality of life itself. Democracy, respect for all human rights and fundamental freedoms, including the right to development, transparent and accountable governance in all sectors of society, as well as effective participation by civil society, are also an essential part of the necessary foundations for the realization of social and people-centered sustainable development.

24. Sustainable development strategies are important mechanisms for enhancing and linking national capacity so as to bring together priorities in social, economic and environmental policies. Hence, special attention must be given to the fulfillment of commitments in the areas set out below, in the framework of an integrated approach towards development, consisting of mutually reinforcing measures to sustain economic growth, as well as to promote social development and environmental protection. Achieving sustainable development cannot be done without greater integration at all policy-making levels and at operational levels, including the lowest administrative levels possible. Economic sectors, such as industry, agriculture, energy, transport and tourism, must take responsibility for the impact of their activities on human well-being and the physical environment. In the context of good governance, properly constructed strategies can enhance prospects for economic growth and employment and at the same time protect the environment. All sectors of society should be involved in their development and implementation, as follows:

(a) By the year 2002, the formulation and elaboration of national strategies for sustainable development which reflect the contributions and responsibilities of all interested parties should be completed in all countries, with assistance provided, as appropriate, through international cooperation, taking into account the special needs of the least developed countries. The efforts of developing countries in effectively implementing national strategies should be supported. Countries which already have national strategies should continue their efforts to enhance and effectively implement them. Assessment of progress achieved and exchange of experience among Governments should be promoted. Local Agenda 21 and other local sustainable development programmes, including youth activities, should also be actively encouraged;

(b) In integrating economic, social and environmental objectives, it is important that a broad package of policy instruments, including regulation, economic instruments, internalization of environmental costs in market prices, environmental and social impact analysis and information, be worked out in the light of country-specific conditions to ensure that integrated approaches are effective and cost-efficient. To this end, a transparent and participatory process should be promoted. This will require the involvement of national legislative assemblies, as well as all actors of civil society, including youth and indigenous people and their communities, to complement the efforts of Governments for sustainable development. In particular, the empowerment and the full and equal participation of women in all spheres of society, including participation in the decision-making process, is central to all efforts to achieve such development;

(c) The implementation of policies aiming at sustainable development, including those contained in Chapter 3 (Combating poverty) and in Chapter 29 (Strengthening the role of workers and their trade unions) of Agenda 21, may enhance the opportunities for job creation, thus helping to achieve the fundamental goal of eradicating poverty.

Enabling international economic climate

25. A mutually supportive balance between the international and the national environment is needed in the pursuit of sustainable development. As a result of globalization, external factors have

become critical in determining the success or failure of developing countries in their national efforts. The gap between
developed and developing countries points to the continued
need for a dynamic and enabling international economic environment supportive of international cooperation, particularly in the
fields of finance, technology transfer, debt and trade, if the
momentum for global progress towards sustainable development
is to be maintained and increased.

26. To foster a dynamic and enabling international economic
    environment favourable to all countries is in the interest of all
    countries. And issues, including environmental issues, that bear
    on the international economic environment can be approached
    effectively only through a constructive dialogue and genuine
    partnership on the basis of mutuality of interests and benefits,
    taking into account that in view of the different contributions to
    global environmental degradation, States have common but
    differentiated responsibilities.

Eradicating poverty

27. Noting the severity of poverty, particularly in developing
    countries, the eradication of poverty is one of the fundamental
    goals of the international community and the entire United
    Nations system, as reflected in commitment 2 of the Copenhagen
    Declaration on Social Development,[7] and is essential for sustain-
    able development. Poverty eradication is thus an overriding
    theme of sustainable development for the coming years. The
    enormity and complexity of the poverty issue could very well
    endanger the social fabric, undermine economic development
    and the environment, and threaten political stability in many
    countries. To achieve poverty eradication, efforts of individual
    Governments and international cooperation and assistance
    should be brought together in a complementary way. Eradication
    of poverty depends on the full integration of people living in
    poverty into economic, social and political life. The empower-
    ment of women is a critical factor for the eradication of poverty.
    Policies that promote such integration to combat poverty, in
    particular policies for providing basic social services and broader
    socio-economic development, are effective as well since enhanc-
    ing the productive capacity of poor people increases both their
    well-being and that of their communities and societies, and
    facilitates their participation in resource conservation and
    environmental protection. The provision of basic social services

and food security in an equitable way is a necessary condition for such integration and empowerment. The 20/20 initiative as referred to in the Programme of Action of the World Summit for Social Development[8] is, among other things, a useful means for such integration. However, the five years since the Rio Conference have witnessed an increase in the number of people living in absolute poverty, particularly in developing countries. In this context, there is an urgent need for the timely and full implementation of all the relevant commitments, agreements and targets already agreed upon since the Rio Conference by the international community, including the United Nations system and international financial institutions. Full implementation of the Programme of Action of the World Summit for Social Development is essential. Priority actions include:

(a) Improving access to sustainable livelihoods, entrepreneurial opportunities and productive resources, including land, water, credit, technical and administrative training, and appropriate technology, with particular efforts to broaden the human and social capital basis of societies to reach the rural poor and the urban informal sector;

(b) Providing universal access to basic social services, including basic education, health care, nutrition, clean water and sanitation;

(c) Progressively developing, in accordance with the financial and administrative capacities of each society, social protection systems to support those who cannot support themselves, either temporarily or permanently; the aim of social integration is to create a "society for all";

(d) Empowering people living in poverty and their organizations by involving them fully in the formulation, implementation and evaluation of strategies and programmes for poverty eradication and community development and by ensuring that these programmes reflect their priorities;

(e) Addressing the disproportionate impact of poverty on women, in particular by removing legislative, policy, administrative and customary barriers to women's equal access to productive resources and services, including access to and control over land and other forms of property, credit, including micro-credit, inheritance, education, information, health care and technology. In this regard, full implementation of the Beijing Platform for Action[9] is essential;

(f)　Interested donors and recipients working together to allocate increased shares of ODA to poverty eradication. The 20/20 initiative is an important principle in this respect, as it is based on a mutual commitment among donors and recipients to increase resources allocated to basic social services;

(g)　Intensifying international cooperation to support measures being taken in developing countries to eradicate poverty, to provide basic social protection and services, and to approach poverty eradication efforts in an integral and multidimensional manner.

Changing consumption and production patterns

28.　Unsustainable patterns of production and consumption, particularly in the industrialized countries, are identified in Agenda 21 as the major cause of continued deterioration of the global environment. While unsustainable patterns in the industrialized countries continue to aggravate the threats to the environment, there remain huge difficulties for developing countries in meeting basic needs such as food, health care, shelter and education for people. All countries should strive to promote sustainable consumption patterns; developed countries should take the lead in achieving sustainable consumption patterns; developing countries should seek to achieve sustainable consumption patterns in the development process, guaranteeing the provision of basic needs to the poor, while avoiding those unsustainable patterns, particularly in industrialized countries, generally recognized as unduly hazardous to the environment, inefficient and wasteful, in the development processes. This requires enhanced technological and other assistance from industrialized countries. In the follow-up of the implementation of Agenda 21, the review of progress made in achieving sustainable consumption patterns should be given high priority.[10] Consistent with Agenda 21, the development and further elaboration of national policies and strategies, particularly in industrialized countries, are needed to encourage changes in unsustainable consumption and production patterns, while strengthening, as appropriate, international approaches and policies that promote sustainable consumption patterns on the basis of the principle of common but differentiated responsibilities, applying the polluter pays principle, encouraging producer responsibility and greater consumer awareness. Eco-efficiency, cost internalization and product policies are also important tools for making consumption and production patterns more sustainable. Actions in this area should focus on:

(a) Promoting measures to internalize environmental costs and benefits in the price of goods and services, while seeking to avoid potential negative effects for market access by developing countries, particularly with a view to encouraging the use of environmentally preferable products and commodities. Governments should consider shifting the burden of taxation on to unsustainable patterns of production and consumption; it is of vital importance to achieve such an internalization of environmental costs. Such tax reforms should include a socially responsible process of reduction and elimination of subsidies to environmentally harmful activities;

(b) Promoting the role of business in shaping more sustainable patterns of consumption by encouraging, as appropriate, voluntary publication of environmental and social assessments of its own activities, taking into account specific country conditions, and by acting as an agent of change in the market, and by virtue of its role as a major consumer of goods and services;

(c) Developing core indicators to monitor critical trends in consumption and production patterns, with industrialized countries taking the lead;

(d) Identifying best practices through evaluations of policy measures with respect to their environmental effectiveness, efficiency and implications for social equity, and disseminating such evaluations;

(e) Taking into account the linkages between urbanization and the environmental and developmental effects of consumption and production patterns in cities, thus promoting more sustainable patterns of urbanization;

(f) Promoting international and national programmes for energy and material efficiency with timetables for their implementation, as appropriate. In this regard, attention should be given to studies that propose to improve the efficiency of resource use, including consideration of a tenfold improvement in resource productivity in industrialized countries in the long term and a possible factor-four increase in industrialized countries in the next two or three decades. Further research is required to study the feasibility of these goals and the practical measures needed for their implementation. Industrialized countries will have a special responsibility and must take the lead in this respect. The Commission on Sustainable Development should consider this initiative in the coming years in exploring policies and measures necessary to implement eco-efficiency and, for this purpose, encourage the relevant bodies to adopt measures aimed at assisting developing countries to improve energy and material efficiency through the promotion of their endogenous capacity-building and economic development with enhanced and effective international support;

(g) Encouraging Governments to take the lead in changing consumption patterns by improving their own environmental performance with action-oriented policies and goals on procurement, the management of public facilities and the further integration of environmental concerns into national policy-making. Governments in developed countries, in particular, should take the lead in this regard;

(h) Encouraging the media, advertising and marketing sectors to help shape sustainable consumption patterns;

(i) Improving the quality of information regarding the environmental impact of products and services and, to that end, encouraging the voluntary and transparent use of eco-labelling;

(j) Promoting measures favouring eco-efficiency; however, developed countries should pay special attention to the needs of developing countries, in particular by encouraging positive impacts, and the need to avoid negative impacts on export opportunities and market access for developing countries and, as appropriate, for countries with economies in transition;

(k) Encouraging the development and strengthening of educational programmes to promote sustainable consumption and production patterns;

(l) Encouraging business and industry to develop and apply environmentally sound technology that should aim not only at increasing competitiveness but also at reducing negative environmental impacts;

(m) Giving balanced consideration to both the demand side and the supply side of the economy in matching environmental concerns and economic factors, which could encourage changes in the behaviour of consumers and producers. A number of policy options should be examined; they include regulatory instruments, economic and social incentives and disincentives, facilities and infrastructure, information, education, and technology development and dissemination.

Making trade and environment mutually supportive

29. In order to accelerate economic growth, poverty eradication and environmental protection, particularly in developing countries, there is a need to establish macroeconomic conditions in both developed and developing countries that favour the development of instruments and structures enabling all countries, in particular developing countries, to benefit from globalization. International cooperation and support for capacity-building in trade, environment and development should be strengthened through renewed system-wide efforts, and with greater responsiveness to sustainable development objectives, by the United Nations, the World Trade Organization (WTO), the Bretton Woods institutions, as well as by national Governments. There should be a balanced and integrated approach to trade and sustainable development, based on a combination of trade liberalization, economic development and environmental protection. Trade obstacles should be removed with a view to contributing to achieving more efficient use of the earth's natural resources in both economic and environmental terms. Trade liberalization should be accompanied by environmental and resource management policies in order to realize its full potential contribution to improved environmental protection and the promotion of sustainable development through the more efficient allocation and use of resources. The multilateral trading system should have the capacity to further integrate environmental considerations and enhance its contribution to sustainable development, without undermining its open, equitable and non-discriminatory character. The special and differential treatment for developing countries, especially the least developed countries, and the other commitments of the Uruguay Round of multilateral trade negotiations should be fully implemented in order to enable those countries to benefit from the international trading system, while conserving the environment. There is a need for continuing the elimination of discriminatory and protectionist practices in international trade relations, which will have the effect of improving access for the

exports of developing countries. This will also facilitate the full
integration of economies in transition into the world economy. In
order to make trade, environment and development mutually
supportive, measures need to be taken to ensure transparency in
the use of trade measures related to the environment, and should
address the root causes of environmental degradation so as not
to result in disguised barriers to trade. Account should be taken
of the fact that environmental standards valid for developed
countries may have unwarranted social and economic costs in
other countries, in particular developing countries. International
cooperation is needed and unilateralism should be avoided. The
following actions are required:

(a) Timely and full implementation of the results of the Uruguay
    Round of Multilateral Trade Negotiations[11] and full use of the
    Comprehensive and Integrated WTO Plan of Action for the
    Least Developed Countries;[12]

(b) Promotion of an open, non-discriminatory, rule-based,
    equitable, secure, transparent and predictable multilateral
    trading system. In this context, effective measures are called
    for to achieve the complete integration of developing
    countries and countries with economies in transition into the
    world economy and the new international trading system. In
    this connection, there is a need to promote the universality
    of WTO and to facilitate the admission to membership in that
    organization in a mutually beneficial way, of developing
    countries and countries with economies in transition apply-
    ing for membership. Actions should be taken to maximize the
    opportunities and to minimize the difficulties of developing
    countries, including the net food-importing ones, especially
    the least developed countries, and of countries with
    economies in transition in adjusting to the changes intro-
    duced by the Uruguay Round. Decisions on further
    liberalization of trade should take into account effects on
    sustainable development and should be consistent with an
    open, rule-based, non-discriminatory, equitable, secure and
    transparent multilateral trading system. The relationship
    between multilateral environmental agreements and the WTO
    rules should be clarified;

(c) Implementation of environmental measures should not result
    in disguised barriers to trade;

(d) Within the framework of Agenda 21, trade rules and environmental principles should interact harmoniously;

(e) Further analysis of the environmental effects of the international transport of goods is warranted;

(f) Cooperation and coordination between the United Nations Conference on Trade and Development (UNCTAD), United Nations Industrial Development Organization (UNIDO), WTO, UNEP and other relevant institutions should be strengthened on various issues, including (i) the role of positive measures in multilateral environmental agreements as part of a package of measures including, in certain cases, trade measures; (ii) the special conditions and needs of small and medium-sized enterprises in the trade and environment interface; (iii) trade and environment issues at the regional and subregional levels, including in the context of regional economic and trade as well as environmental agreements;

(g) Cooperation and coordination between UNCTAD and other relevant bodies within their existing respective mandates should be enhanced, inter alia, on environment and sustainable development issues. Without prejudice to the clear understanding in WTO that future negotiations, if any, regarding a multilateral agreement on investment will take place only after an explicit consensus decision, future agreements on investments should take into account the objectives of sustainable development and, when developing countries are parties to these agreements, special attention should be given to their needs for investment;

(h) National Governments should make every effort to ensure policy coordination on trade, environment and development at the national level in support of sustainable development;

(i) There is a need for the WTO, UNEP and UNCTAD to consider ways to make trade and environment mutually supportive, including through due respect to the objectives and principles of the multilateral trading system and to the provisions of multilateral environmental agreements. Such considerations should be consistent with an open, rule-based, non-discriminatory, equitable, secure and transparent multilateral trading system.

Population

30. The impact of the relationship between economic growth, poverty, employment, environment and sustainable development

has become a major concern. There is a need to recognize the critical linkages between demographic trends and factors and sustainable development. The current decline in population growth rates must be further promoted through national and international policies that promote economic development, social development, environmental protection, poverty eradication, particularly the further expansion of basic education, with full and equal access for girls and women, and health care, including reproductive health care, including both family planning and sexual health, consistent with the report of the International Conference on Population and Development.[13]

Health

31. The goals of sustainable development cannot be achieved when a high proportion of the population is afflicted with debilitating illnesses. An overriding goal for the future is to implement the Health for All strategy[14] and to enable all people, particularly the world's poor, to achieve a higher level of health and well-being, and to improve their economic productivity and social potential. Protecting children from environmental health threats and infectious disease is particularly urgent since children are more susceptible than adults to those threats. Top priority should be attached to supporting the efforts of countries, particularly developing countries, and international organizations to eradicate the major infectious diseases, especially malaria, which is on the increase, to improve and expand basic health and sanitation services, and to provide safe drinking water. It is also important to reduce indigenous cases of vaccine-preventable diseases through the promotion of widespread immunization programmes, promote accelerated research and vaccine development and reduce the transmission of other major infectious diseases, such as dengue fever, tuberculosis and HIV/AIDS. Given the severe and irreversible health effects of lead poisoning, particularly on children, it is important to accelerate the process of eliminating unsafe uses of lead, including the use of lead in gasoline worldwide, in light of country-specific conditions and with enhanced international support and assistance to developing countries, particularly through the timely provision of technical and financial assistance and the promotion of endogenous capacity-building. Strategies at the regional, national and local levels for reducing the potential risk due to ambient and

indoor air pollution should be developed, bearing in mind their serious impacts on human health including strategies to make parents, families and communities aware of the adverse environmental health impacts of tobacco. The clear linkage between health and the environment needs to be emphasized and the lack of information on the impact of environmental pollution on health should be addressed. Health issues should be fully integrated into national and subnational sustainable development plans and should be incorporated into project and programme development as a component of environmental impact assessments. Important to efforts at national levels is international cooperation in disease prevention, early warning, surveillance, reporting, training and research, and treatment.

Sustainable human settlements

32. Sustainable human settlements development is essential to sustainable development. The need to intensify efforts and cooperation to improve living conditions in the cities, towns, villages and rural areas throughout the world is recognized. Approximately half the world's population already lives in urban settlements, and by early in the next century the majority – more than 5 billion people – will be urban residents. Urban problems are concerns common to both developed and developing countries, although urbanization is occurring most rapidly in developing countries. Urbanization creates both challenges and opportunities. Global urbanization is a cross-sectoral phenomenon which has an impact on all aspects of sustainable development. Urgent action is needed to implement fully the commitments made at the United Nations Conference on Human Settlements (Habitat II) consistent with its report,[15] and in Agenda 21. New and additional financial resources from various sources are necessary to achieve the goals of adequate shelter for all and sustainable human settlements development in an urbanizing world. Transfer of expertise and technology, · capacity-building, decentralization of authority through, inter alia, strengthening of local capacity and private–public partnerships to improve the provision and environmentally sound management of infrastructure and social services should be accelerated to achieve more sustainable human settlements development. Local Agenda 21 programmes should also be actively encouraged. · Global targets could be established by the Commission on

Sustainable Development to promote Local Agenda 21 campaigns and to deal with obstacles to Local Agenda 21 initiatives.

2. SECTORS AND ISSUES

33. The present section identifies a number of specific areas that are of widespread concern since failure to reverse current trends in these areas, notably in resource degradation, will have potentially disastrous effects on social and economic development, on human health and on environmental protection for all countries, particularly developing countries. All sectors covered by Agenda 21 are equally important and thus deserve attention by the international community on an equal footing. The need for integration is important in all sectors, including the areas of energy and transport because of the adverse effects that developments in those areas can have on human health and ecosystems; the areas of agriculture and water use, where inadequate land-use planning, poor water management and inappropriate technology can result in the degradation of natural resources and human impoverishment and where drought and desertification can result in land degradation and soil loss; and the area of management of marine resources, where competitive overexploitation can damage the resource base, food supplies and the livelihood of fishing communities, as well as the environment. The recommendations made in each of the sectors take into account the need for international cooperation in support of national efforts, within the context of the principles of UNCED, including, inter alia, the principle of common but differentiated responsibilities. It is likewise understood that these recommendations do not in any way prejudice the work accomplished under legally binding conventions, where they exist, concerning these sectors.

Freshwater

34. Water resources are essential for satisfying basic human needs, health and food production, and the preservation of ecosystems, as well as for economic and social development in general. It is a matter of urgent concern that more than one fifth of all people still do not have access to safe drinking water and more than one half of humanity lacks adequate sanitation. From the perspective of developing countries, freshwater is a priority and a basic need, especially taking into account that in many developing countries freshwater is not readily available for all segments of the popula-

tion, inter alia, owing to lack of adequate infrastructure and capacity, water scarcity, and technical and financial constraints. Moreover, freshwater is also crucial for developing countries to satisfy the basic needs of their population in the areas of agricultural irrigation, industrial development, hydroelectric generation, and so forth. In view of the growing demands on water, which is a finite resource, water will become a major limiting factor in socio-economic development unless early action is taken. There is growing concern at the increasing stress on water supplies caused by unsustainable use patterns, affecting both water quality and quantity, and the widespread lack of access to safe water supply and suitable sanitation in many developing countries. Because the commitments of the International Drinking Water Supply and Sanitation Decade of the 1980s have not been fully met, there is still a need to ensure the optimal use and protection of all freshwater resources, so that the needs, including the availability of safe drinking water and sanitation, of everyone on this planet can be met. This calls for the highest priority to be given to the serious freshwater problems facing many regions, especially in the developing world. There is an urgent need to:

(a) Assign high priority, in accordance with specific national needs and conditions, to the formulation and implementation of policies and programmes for integrated watershed management, including issues related to pollution and waste, the interrelationship between water and land, including mountains, forests, upstream and downstream users, estuarine environments, biodiversity and the preservation of aquatic ecosystems, wetlands, climate and land degradation and desertification, recognizing that subnational, national and regional approaches to freshwater protection and consumption following a watershed basin or river basin approach offer a useful model for the protection of freshwater supplies;

(b) Strengthen regional and international cooperation for technological transfer and the financing of integrated water resources programmes and projects, in particular those designed to increase access to safe water supply and sanitation;

(c) Ensure the continued participation of local communities, and women in particular, in the management of water resources development and use;

(d) Provide an enabling national and international environment that encourages investments from public and private sources to improve water supply and sanitation services, especially in fast-growing urban and peri-urban areas, as well as in poor rural communities in developing countries. Adopt and implement commitments by the international community to support the efforts to assist developing countries achieve access to safe drinking water and sanitation for all;

(e) Recognize water as a social and economic good with a vital role in the satisfaction of basic human needs, food security, poverty alleviation and the protection of ecosystems. Economic valuation of water should be seen within the context of its social and economic implications, reflecting the importance of meeting basic needs. Consideration should be given to the gradual implementation of pricing policies that are geared towards cost recovery and the equitable and efficient allocation of water, including the promotion of water conservation, in developed countries; such policies could also be considered in developing countries when they reach an appropriate stage in their development, so as to promote the harmonious management and development of scarce water resources and generate financial resources for investment in new water supply and treatment facilities. Such strategies should also include programmes assigned to minimize wasteful consumption of water;

(f) Strengthen the capability of Governments and international institutions to collect and manage information, including scientific, social and environmental data, in order to facilitate the integrated assessment and management of water resources, and foster regional and international cooperation for information dissemination and exchange through cooperative approaches among United Nations institutions, including UNEP, and centres for environmental excellence. In this regard, technical assistance to developing countries will continue to be important;

(g) Give support by the international community to the efforts and limited resources of developing countries to shift to higher-value, less water-intensive modes of agricultural and industrial production and to develop the educational and informational infrastructure necessary to improve the skills of the labour force required for the economic transformation that needs to take place if use of freshwater resources is to be sustainable. International support for the integrated development of water resources in developing countries, appropriate innovative initiatives and approaches at the bilateral and regional levels are also required;

(h) Encourage watercourse States to develop international watercourses with a view to attaining sustainable utilization and appropriate protection thereof and benefits therefrom, taking into account the interests of the watercourse States concerned.

35. Considering the urgent need for action in the field of freshwater, and building on existing principles and instruments, arrangements, programmes of action and customary uses of water, Governments call for a dialogue under the aegis of the Commission on Sustainable Development, beginning at its sixth session, aimed at building a consensus on the necessary actions, and in particular, on the means of implementation and on tangible results, in order to consider initiating a strategic approach for the implementation of all aspects of the sustainable use of freshwater for social and economic purposes, including, inter alia, safe drinking water and sanitation, water for irrigation, recycling, and wastewater management, and the important role freshwater plays in natural ecosystems. This intergovernmental process will be fully fruitful only if there is a proved commitment by the international community for the provision of new and additional financial resources for the goals of this initiative.

Oceans and seas

36. Progress has been achieved since UNCED in the negotiation of agreements and voluntary instruments for improving the conservation and management of fishery resources and for the protection of the marine environment. Furthermore, progress has been made in the conservation and management of specific fishery stocks for securing the sustainable utilization of these resources. Despite this, the decline of many fish stocks, high levels of discards, and rising marine pollution continue. Governments should take full advantage of the challenge and opportunity presented by the International Year of the Ocean in 1998. There is a need to continue to improve decision-making at the national, regional and global levels. To address the need for improving global decision-making on the marine environment,

there is an urgent need for Governments to implement decision 4/15 of the Commission on Sustainable Development,[16] in which the Commission, inter alia, called for periodic intergovernmental reviews by the Commission of all aspects of the marine environment and its related issues, as described in chapter 17 of Agenda 21, for which the overall legal framework is provided by UNCLOS. There is a need for concerted action by all countries and for improved cooperation to assist developing countries in implementing the relevant agreements and instruments in order to participate effectively in the sustainable use, conservation and management of their fishery resources, as provided for in UNCLOS and other international legal instruments and to achieve integrated coastal zone management. In that context, there is an urgent need for:

(a) All Governments to ratify or to accede to the relevant agreements as soon as possible and to implement effectively such agreements as well as relevant voluntary instruments;

(b) All Governments to implement General Assembly resolution 51/189 of 16 December 1996, including the strengthening of institutional links to be established between the relevant intergovernmental mechanisms involved in the development and implementation of integrated coastal zone management. Following progress on UNCLOS, and bearing in mind principle 13 of the Rio Declaration on Environment and Development, there is a need to strengthen the implementation of existing international and regional agreements on marine pollution, with a view in particular to better contingency planning, response, and liability and compensation mechanisms;

(c) Better identification of priorities for action at the global level to promote the conservation and sustainable use of the marine environment, as well as better means for integrating such action;

(d) Further international cooperation to support the strengthening, where needed, of regional and subregional agreements for the protection and sustainable use of the oceans and seas;

(e) Governments to prevent or eliminate overfishing and excess fishing capacity through the adoption of management measures and mechanisms to ensure the sustainable management and utilization of fishery resources and to undertake programmes of work to achieve the reduction and elimination of wasteful fishing practices, wherever they may occur, especially in relation to large-scale industrialized fishing. The emphasis given by the Commission on Sustainable Development at its fourth session to the importance of effective conservation and management of fish stocks, and in particular to eliminating overfishing, in order to identify specific steps at the national or regional levels to prevent or eliminate excess fishing capacity, will need to be carried forward in all appropriate international forums including, in particular, the Committee on Fisheries of the Food and Agriculture Organization of the United Nations (FAO);

(f) Governments to consider the positive and negative impact of subsidies on the conservation and management of fisheries through national, regional and appropriate international organizations and, based on these analyses, to consider appropriate action;

(g) Governments to take actions, individually and through their participation in competent global and regional forums, to improve the quality and quantity of scientific data as a basis for effective decisions related to the protection of the marine environment and the conservation and management of marine living resources; in this regard, greater international cooperation is required to assist developing countries, in particular small island developing States, to operationalize data networks and clearing houses for information-sharing on oceans. In this context, particular emphasis must be placed on the collection of biological and other fisheries-related information and the resources for its collation, analysis and dissemination.

Forests

37. The management, conservation and sustainable development of all types of forests is a crucial factor in economic and social development, in environmental protection and in the planet's life support system. Forests are one of the major reservoirs of biological diversity; they act as carbon sinks and reservoirs; and they are a significant source of renewable energy, particularly in the least developed countries. Forests are an integral part of sustainable development and are essential to many indigenous people and other forest-dependent people embodying traditional lifestyles, forest owners and local communities, many of whom possess

important traditional forest-related knowledge.

38. Since the adoption of the Forest Principles at the Rio Conference, tangible progress has been made in sustainable forest management at the national, subregional, regional and international levels and in the promotion of international cooperation on forests. The proposals for action contained in the report of the Ad Hoc Intergovernmental Panel on Forests (IPF) (E/CN.17/1997/12), which were endorsed by the Commission on Sustainable Development at its fifth session,[17] represent significant progress and consensus on a wide range of forest issues.

39. To maintain the momentum generated by the IPF process and to facilitate and encourage the holistic, integrated and balanced intergovernmental policy dialogue on all types of forests in the future, which continues to be an open, transparent and participatory process, requires a long-term political commitment to sustainable forest management worldwide. Against this background, there is an urgent need for:

    (a) Countries and international organizations and institutions to implement the proposals for action agreed by the Panel, in an expeditious and effective manner, and in collaboration and through effective partnership with all interested parties, including major groups, in particular indigenous people and local communities;

    (b) Countries to develop national forest programmes in accordance with their respective national conditions, objectives and priorities;

    (c) Enhanced international cooperation to implement the Panel's proposals for action directed towards the management, conservation and sustainable development of all types of forests, including provision for financial resources, capacity-building, research and the transfer of technology;

    (d) Further clarification of all issues arising from the programme elements of the IPF process;

    (e) International institutions and organizations to continue their work and to undertake further coordination and explore means for collaboration in the informal, high-level Inter-Agency Task Force on Forests, focusing on the implementation of the Panel's proposals for action, in accordance with their respective mandates and comparative advantage;

    (f) Countries to provide consistent guidance to the governing bodies of relevant international institutions and instruments to take efficient and effective measures, as well as to coordinate their forest-related work at all levels, in incorporating the Panel's proposals for action into their work programmes and under existing agreements and arrangements.

40. To help achieve this, it is decided to continue the intergovernmental policy dialogue on forests through the establishment of an ad hoc, open-ended Intergovernmental Forum on Forests under the aegis of the Commission on Sustainable Development to work in an open, transparent and participatory manner, with a focused and time-limited mandate, charged with, inter alia:

    (a) Promoting and facilitating the implementation of the Panel's proposals for action;

    (b) Reviewing, monitoring and reporting on progress in the management, conservation and sustainable development of all types of forests;

    (c) Considering matters left pending on the programme elements of the IPF, in particular trade and environment in relation to forest products and services, transfer of technology and the need for financial resources.

    The Forum should also identify the possible elements of and work towards consensus for international arrangements and mechanisms, for example a legally binding instrument. The Forum will report on its work to the Commission for Sustainable Development in 1999. Based on that report, and depending on the decision of the Commission at its eighth session, the Forum will engage in further action on establishing an intergovernmental negotiation process on new arrangements and mechanisms or a legally binding instrument on all types of forests.

41. The Forum should convene as soon as possible to further elaborate its terms of reference and decide on organizational matters. It should be serviced by a small secretariat within the Department of Policy Coordination and Sustainable Development supported by voluntary extrabudgetary contributions from governments and international organizations.

Energy

42. Energy is essential to economic and social development and improved quality of life. However, sustainable patterns of production, distribution and use of energy are crucial. Fossil fuels (coal, oil and natural gas) will continue to dominate the energy supply situation for many years to come in most developed and develop-

ing countries. What is required then is to reduce the environmental impact of their continued development, and to reduce local health hazards and environmental pollution through enhanced international cooperation notably in the provision of concessional finance for capacity development and transfer of the relevant technology, and through appropriate national action.

43. In developing countries sharp increases in energy services are required to improve the standard of living of their growing populations. The increase in the level of energy services would have a beneficial impact on poverty eradication by increasing employment opportunities and improving transportation, health and education. Many developing countries, in particular the least developed, face the urgent need to provide adequate modern energy services, especially to billions of people in rural areas. This requires significant financial, human and technical resources and a broad-based mix of energy sources.

44. The objectives envisaged in this section should reflect the need for equity, adequate energy supplies and increasing energy consumption in developing countries and should take into account the situation of countries that are highly dependent on income generated from the production, processing and export, and/or consumption of fossil fuels and that have serious difficulties in switching to alternative sources of energy, and the situation of countries highly vulnerable to adverse effects of climate change.

45. Advances towards sustainable energy use are taking place and all parties can benefit from progress made in other countries. It is also necessary to ensure international cooperation for promoting energy conservation and improvement of energy efficiency, the use of renewable energy and research, and the development and dissemination of innovative energy-related technology.

46. Therefore there is a need for:

(a) A movement towards sustainable patterns of production, distribution and use of energy. To advance this work at the intergovernmental level, the Commission on Sustainable Development will discuss energy issues at its ninth session. Noting the vital role of energy in the continuation of sustained economic growth, especially for developing countries, be they importers or suppliers of energy, and recognizing the complexities and interdependencies inherent in addressing energy issues within the context of sustainable development, preparations for this session should be initiated at the seventh session and should utilize an open-ended intergovernmental group of experts on energy and sustainable development to be held in conjunction with inter-sessional meetings of the eighth and ninth sessions of the Commission. In line with the objectives of Agenda 21, the ninth session of the Commission should contribute to a sustainable energy future for all;

(b) Evolving concrete measures to strengthen international cooperation in order to assist developing countries in their domestic efforts to provide adequate modern energy services, especially electricity, to all sections of their population, particularly in rural areas, in an environmentally sound manner;

(c) Countries, bearing in mind the specific needs and priorities of developing countries, to promote policies and plans that take into account the economic, social and environmental aspects of the production, distribution and use of energy, including the use of lower pollutant sources of energy such as natural gas;

(d) Evolving commitments for the transfer of relevant technology, including time-bound commitments, as appropriate, to developing countries and economies in transition so as to enable them to increase the use of renewable energy sources and cleaner fossil fuels and to improve efficiency in energy production, distribution and use. Countries need to systematically increase the use of renewable energy sources according to their specific social, economic, natural, geographical and climatic conditions and cleaner fuel technologies, including fossil fuel technologies, and to improve efficiency in energy production, distribution and use and in other industrial production processes that are intensive users of energy;

(e) Promoting efforts in research on and development and use of renewable energy technologies at the international and national levels;

(f) In the context of fossil fuels, encouraging further research, development, and the application and transfer of technology of a cleaner and more efficient nature, through effective international support;

(g) Encouraging Governments and the private sector to consider appropriate ways to gradually promote environmental cost

internalization so as to achieve more sustainable use of
energy, taking fully into account the economic, social and
environmental conditions of all countries, in particular developing countries. In this regard, the international community
should cooperate to minimize the possible adverse impacts
on the development process of developing countries resulting from the implementation of those policies and measures.
There is also a need to encourage the reduction and the
gradual elimination of subsidies for energy production and
consumption that inhibit sustainable development. Such
policies should take fully into account the specific needs and
conditions of developing countries, particularly least developed countries, as reflected in the special and differential
treatment accorded them in the Uruguay Round of
Multilateral Trade Negotiations Agreement on Subsidies and
Countervailing Measures;[18]

(h) Encouraging better coordination on the issue of energy
within the United Nations system, under the guidance of the
General Assembly and taking into account the coordinating
role of ECOSOC.

Transport

47. The transport sector and mobility in general have an essential
and positive role to play in economic and social development,
and transportation needs will undoubtedly increase. Over the
next 20 years, transportation is expected to be the major driving
force behind a growing world demand for energy. The transport
sector is the largest end-user of energy in developed countries
and the fastest growing one in most developing countries.
Current patterns of transportation with their dominant patterns
of energy use are not sustainable and on present trends may
compound the environmental problems the world is facing and
the health impacts referred to in paragraph 25 above. There is a
need for:

(a) The promotion of integrated transport policies that consider
alternative approaches to meeting commercial and private
mobility needs and improve performance in the transport
sector at the national, regional and global levels, and particularly a need to encourage international cooperation in the
transfer of environmentally sound technologies in the transport sector and implementation of appropriate training

programmes in accordance with national programmes and
priorities;

(b) The integration of land use and urban, peri-urban and rural
transport planning, taking into account the need to protect
ecosystems;

(c) The adoption and promotion, as appropriate, of measures to
mitigate the negative impact of transportation on the environment, including measures to improve efficiency in the
transportation sector;

(d) The use of a broad spectrum of policy instruments to
improve energy efficiency and efficiency standards in transportation and related sectors;

(e) The continuation of studies in the appropriate fora, including
the International Civil Aviation Organization (ICAO), on the
use of economic instruments for the mitigation of the
negative environmental impact of aviation in the context of
sustainable development;

(f) Accelerating the phasing-out of the use of leaded gasoline as
soon as possible, in pursuit of the objectives of reducing the
severe health impacts of human exposure to lead. In this
regard technological and economic assistance should
continue to be provided to developing countries in order to
enable them to make such a transition;

(g) The promotion of voluntary guidelines for environmentally
friendly transport, and actions for reducing vehicle emissions
of carbon dioxide, carbon monoxide, nitrogen oxides, partic-
ulate matter and volatile organic compounds, as soon as
possible;

(h) Partnerships at the national level, involving Governments,
local authorities, non-governmental organizations and the
private sector, for strengthening transport infrastructures and
developing innovative mass transport schemes.

Atmosphere

48. Ensuring that the global climate and atmosphere is not further
damaged with irreversible consequences for future generations
requires political will and concerted efforts by the international
community in accordance with the principles enshrined in the
United Nations Framework Convention on Climate Change.
Under the Convention, some first steps have been taken to deal
with the global problem of climate change. Despite the adoption

of the Convention, the emission and concentration of green-house gases (GHGs) continue to rise, even as scientific evidence assembled by the Intergovernmental Panel on Climate Change (IPCC) and other relevant bodies continues to diminish the uncertainties and points ever more strongly to the severe risk of global climate change. So far, insufficient progress has been made by many developed countries in meeting their aim to return GHG emissions to 1990 levels by the year 2000. It is recognized as one critical element of the Berlin Mandate[19] that the commitments under article 4, paragraph 2 (a) and (b) of the Convention are inadequate and that therefore there is a need to strengthen these commitments. It is most important that the Conference of Parties to the Convention, at its third session, to be held at Kyoto, Japan later in 1997, adopt a protocol or other legal instrument that fully encompasses the Berlin Mandate. The Geneva Ministerial Declaration[20] which was noted without formal adoption, but which received majority support among ministers and other heads of delegation attending the second session of the Conference of the Parties, also called for, inter alia, the acceleration of negotiations on the text of a legally binding protocol or other legal instrument.

49. At the nineteenth special session of the General Assembly, the international community confirmed its recognition of the problem of climate change as one of the biggest challenges facing the world in the next century. The leaders of many countries underlined the importance of this in their addresses to the Assembly, and outlined the actions they have in hand both in their own countries and internationally to respond.

50. The ultimate goal which all countries share is to achieve stabilization of greenhouse gas concentrations in the atmosphere at a level that would prevent dangerous anthropogenic interference with the climate system. This requires efficient and cost-effective policies and measures that will be sufficient to result in a significant reduction in emissions. At this session, countries reviewed the state of preparations for the third session of the Conference of Parties of the Framework Convention of Climate Change in Kyoto. All are agreed that it is vital that there should be a satisfactory result.

51. The positions of many countries for these negotiations are still evolving, and it was agreed that it would not be appropriate to seek to predetermine the results, although useful interactions on evolving positions took place.

52. There is already widespread but not universal agreement that it will be necessary to consider legally binding, meaningful, realistic and equitable targets for annex I countries that will result in significant reductions in greenhouse gas emissions within specified time frames, such as 2005, 2010 and 2020. In addition to establishing targets, there is also widespread agreement that it will be necessary to consider ways and means for achieving them and to take into account the economic, adverse environmental and other effects of such response measures on all countries, particularly developing countries.

53. International cooperation in the implementation of chapter 9 of Agenda 21, in particular in the transfer of technology to and capacity-building in developing countries, is also essential to promote the effective implementation of the United Nations Framework Convention on Climate Change.

54. There is also a need to strengthen systematic observational networks to identify the possible onset and distribution of climate change and assess potential impacts, particularly at the regional level.

55. The ozone layer continues to be severely depleted and the Montreal Protocol[21] needs to be strengthened. The Copenhagen Amendment to the Protocol needs to be ratified. The recent successful conclusion of the replenishment negotiations of the Montreal Protocol Multilateral Fund is welcomed. This has made available funds for, among other things, earlier phase-out of ozone-depleting substances, including methyl bromide, in developing countries. Future replenishment should also be adequate to ensure timely implementation of the Montreal Protocol. An increased focus on capacity-building programmes in developing countries within multilateral funds is also needed, as well as the implementation of effective measures against illegal trade in ozone-depleting substances.

56. Rising levels of transboundary air pollution should be countered, including through appropriate regional cooperation to reduce pollution levels.

Toxic chemicals

57. The sound management of chemicals is essential to sustainable development and is a fundamental underpinning human health and environmental protection. All those responsible for chemicals, throughout their life cycle, bear the responsibility for

achieving this. Substantial progress on the sound management of
chemicals has been made since UNCED, in particular through the
establishment of the Intergovernmental Forum on Chemical
Safety (IFCS) and the Inter-Organizational Programme for the
Sound Management of Chemicals (IOMC). In addition, domestic
regulations have been complemented by the Code of Ethics on
the International Trade in Chemicals and by voluntary industry
initiatives, such as Responsible Care. Despite substantial
progress, a number of chemicals continue to pose significant
threats to local, regional and global ecosystems and to human
health. Since UNCED, there has been an increased understanding
of the serious damage that certain toxic chemicals can cause to
human health and the environment. Much remains to be done
and the environmentally sound management of chemicals should
continue to be an important issue well beyond 2000. Particular
attention should also be given to cooperation in the development
and transfer of technology of safe substitutes and in the development of capacity for the production of such substitutes. The
decision on the sound management of chemicals adopted by the
Governing Council of UNEP at its nineteenth session should be
implemented in accordance with the agreed timetables for
negotiations on the prior informed consent (PIC) and persistent
organic pollutants (POPs) conventions. It is noted that inorganic
chemicals possess roles and behaviour that are distinct from
organic chemicals.

Hazardous wastes

58. Substantial progress has been made with the implementation of
the Basel Convention,[22] the Bamako Convention,[23] the Fourth
Lome' Convention and other regional Conventions, although
more remains to be done. There are important initiatives aimed
at promoting the environmentally sound management of
hazardous wastes under the Basel Convention, including (a)
activities undertaken to prevent illegal traffic in hazardous
wastes; (b) the establishment of regional centres for training and
technology transfer regarding hazardous waste minimization and
management; and (c) the treatment and disposal of hazardous
wastes as close as possible to their source of origin. These initiatives should be further developed. It is also important and urgent
that work under the Basel Convention be completed to define
which hazardous wastes are controlled under the Convention

and to negotiate, adopt and implement a protocol on liability and
compensation for damage resulting from transboundary
movements and disposal of hazardous wastes. Land contaminated
by the disposal of hazardous wastes needs to be identified and
remedial actions put in hand. Integrated management solutions
are also required to minimize urban and industrial waste generation and to promote recycling and reuse.

Radioactive wastes

59. Radioactive wastes can have very serious environmental and
human health impacts over long periods of time. Therefore, it is
essential that they be managed in a safe and responsible way. The
storage, transportation, transboundary movement and disposal of
radioactive wastes should be guided by all the principles of the
Rio Declaration, and by Agenda 21. States which generate
radioactive wastes have a responsibility to ensure their safe
storage and disposal. In general, radioactive wastes should be
disposed of in the territory of the State in which they are generated as far as is compatible with the safety of the management of
such material. Each country has a responsibility to ensure that
radioactive wastes that fall within its jurisdiction are managed
properly in accordance with internationally accepted principles,
taking fully into account any transboundary effects. The international community should make all efforts to prohibit the export
of radioactive wastes to those countries that do not have appropriate waste treatment and storage facilities. The international
community also recognizes that regional arrangements or jointly
used facilities might be appropriate for the disposal of such
wastes in certain circumstances. The management[24] of radioactive
wastes should be undertaken in a manner consistent with international law, including the provisions of relevant international
and regional conventions and with internationally accepted
standards. It is important to intensify safety measures with regard
to radioactive wastes. States, in cooperation with relevant international organizations, where appropriate, should not promote or
allow the storage or disposal of high-level, intermediate-level and
low-level radioactive wastes near the marine environment unless
they determine that scientific evidence, consistent with the
applicable internationally agreed principles and guidelines,
shows that such storage or disposal poses no unacceptable risk to
people or the marine environment and does not interfere with

other legitimate uses of the sea. In the process of the consideration of that evidence, appropriate application of the
precautionary approach principle should be made. Further action
is needed by the international community to address the need for
enhancing awareness of the importance of safe management of
radioactive wastes, and to ensure the prevention of incidents and
accidents involving the uncontrolled release of such wastes.

60. One of the main recommendations of Agenda 21 and of the
Commission on Sustainable Development at its second session in
this area was to support the ongoing efforts of the International
Atomic Energy Agency (IAEA), the International Maritime
Organization (IMO) and other relevant international organizations. The Joint Convention on the Safety of Spent Fuel
Management and on the Safety of Radioactive Waste Management
currently being negotiated under the auspices of IAEA is now
close to completion. It will provide a comprehensive codification
of international law and a guide to best practices in this area. It
will rightly be based on all the principles of best practice for this
subject that have evolved in the international community, including the principle that, in general, radioactive wastes should be
disposed of in the State in which they were generated as far as is
compatible with the safety of the management of such material.
Governments should finalize this text and are urged to ratify and
implement it as soon as possible so as to further improve practice
and strengthen safety in this area. Transportation of irradiated
nuclear fuel (INF) and high-level waste by sea should be guided
by the INF Code, which should be considered for development
into a mandatory instrument. The issue of potential transboundary environmental effects of activities related to the management
of radioactive wastes and the question of prior notification,
relevant information and consultation with States that could
potentially be affected by such effects, should be further
addressed within the appropriate forums.

61. Increased global and regional cooperation, including exchange of
information and experience and transfer of appropriate technologies, is needed to improve the management of radioactive wastes.
There is a need to support the clean-up of sites contaminated as
a result of all types of nuclear activity and to conduct health
studies in the regions around those sites, as appropriate, with a
view to identifying where health treatment may be needed and
should be provided. Technical assistance should be provided to

developing countries, recognizing the special needs of small island developing States in particular, to enable them to develop or improve procedures for the management and safe disposal of radioactive wastes deriving from the use of radionuclides in medicine, research and industry.

Land and sustainable agriculture

62. Land degradation and soil loss threaten the livelihood of millions of people and future food security, with implications for water resources and the conservation of biodiversity. There is an urgent need to define ways to combat or reverse the worldwide accelerating trend of soil degradation, using an ecosystem approach, taking into account the needs of populations living in mountain ecosystems and recognizing the multiple functions of agriculture. The greatest challenge for humanity is to protect and sustainably manage the natural resource base on which food and fibre production depend, while feeding and housing a population that is still growing. The international community has recognized the need for an integrated approach to the protection and sustainable management of land and soil resources, as stated in decision III/11 of the Conference of the Parties to the Convention on Biological Diversity,[25] including identification of land degradation, which involves all interested parties at the local as well as the national level, including farmers, small-scale food producers, indigenous people and their communities, non-governmental organizations and, in particular, women, who have a vital role in rural communities. This should include action to ensure secure land tenure and access to land, credits and training as well as the removal of obstacles that inhibit farmers, especially small-scale farmers and peasants, from investing in and improving their lands and farms.

63. It remains essential to continue efforts for the eradication of poverty through, inter alia, capacity-building to reinforce local food systems, improving food security and providing adequate nutrition for the more than 800 million undernourished people in the world, located mainly in developing countries.
Governments should formulate policies that promote sustainable agriculture as well as productivity and profitability.
Comprehensive rural policies are required to improve access to land, combat poverty, create employment and reduce rural emigration. In accordance with commitments agreed to in the

Rome Declaration on World Food Security and the World Food
Summit Plan of Action, adopted by the World Food Summit
(Rome, 13–17 November 1996),[26] sustainable food security
among both the urban and the rural poor should be a policy
priority, and developed countries and the international community should provide assistance to developing countries to this
end. To meet those objectives, Governments should attach high
priority to implementing the commitments of the Rome
Declaration and Plan of Action, especially the call for a minimum
target of halving the number of undernourished people in the
world by the year 2015. Governments and international organizations are encouraged to implement the Global Plan of Action for
the Conservation and Sustainable Utilization of Plant Genetic
Resources for Food and Agriculture as adopted by the
International Technical Conference on Plant Genetic Resources
(Leipzig, Germany, 17–23 June 1996). At the sixth session of the
Commission on Sustainable Development, in 1998, the issues of
sustainable agriculture and land use should be considered in
relation to freshwater. The challenge for agricultural research is
to increase yields on all farmlands while protecting and conserving the natural resource base. The international community and
Governments must continue or increase investments in agricultural research because it can take years or decades to develop
new lines of research and bring those research findings into
sustainable practice on the land. Developing countries, particularly those with high population densities, will need international
cooperation to gain access to the results of such research and to
technology aimed at improving agricultural productivity in
limited spaces. More generally, international cooperation continues to be needed to assist developing countries in many other
aspects of basic requirements of agriculture. There is a need to
support the continuation of the reform process in conformity
with the Uruguay Round Agreement, particularly Article 20 of the
Agreement on Agriculture, and to fully implement the WTO
Decision on Measures Concerning the Possible Negative Effects of
the Reform Programme on Least-Developed and Net
Food-Importing Developing Countries.

Desertification and drought

64. Governments are urged to conclude – by signing and ratifying,
    accepting, approving and/or acceding to – and to implement the

United Nations Convention to Combat Desertification in Those Countries Experiencing Serious Drought and/or Desertification, Particularly in Africa, which entered into force on 26 December 1996, as soon as possible, and to support and actively participate in the first session of the Conference of the Parties to the Convention, which will be held in Rome in September 1997.

65. The international community is urged to recognize the vital importance and necessity of international cooperation and partnership in combating desertification and mitigating the effects of drought. In order to increase the effectiveness and efficiency of existing financial mechanisms, the international community, in particular developed countries, should therefore support the global mechanism that would indeed have the capacity to promote actions leading to the mobilization and channeling of substantial resources for advancing the implementation of the Convention and its regional annexes, and to contribute to the eradication of poverty, which is one of the principal consequences of desertification and drought in the majority of affected countries. Another view considers that the international community, in particular developed countries, should provide new and additional resources towards the same ends. The transfer to developing countries of environmentally sound, economically viable and socially acceptable technologies relevant to combating desertification and/or mitigating the effects of drought, with a view to contributing to the achievement of sustainable development in affected areas, should be undertaken without delay on mutually agreed terms.

Biodiversity

66. There remains an urgent need for the conservation and sustainable use of biological diversity and the fair and equitable sharing of benefits arising from the utilization of components of genetic resources. The threat to biodiversity stems mainly from habitat destruction, over-harvesting, pollution and the inappropriate introduction of foreign plants and animals. There is an urgent need for Governments and the international community, with the support of relevant international institutions, as appropriate, to:
    (a) Take decisive action to conserve and maintain genes, species and ecosystems with a view to promoting sustainable management of biological diversity;
    (b) Ratify the Convention on Biological Diversity and implement

it fully and effectively together with the decisions of the Conference of the Parties, including recommendations on agricultural biological diversity and the Jakarta Mandate on Marine and Coastal Biological Diversity, and pursue urgently other tasks identified by the Conference of the Parties at its third meeting under the work programme on terrestrial biological diversity,[27] within the context of the ecosystems approach adopted in the Convention;

(c) Undertake concrete actions for the fair and equitable sharing of the benefits arising from the utilization of genetic resources, consistent with the provisions of the Convention and the decisions of the Conference of the Parties on, inter alia, access to genetic resources and handling of biotechnology and its benefits;

(d) Pay further attention to the provision of new and additional financial resources for the implementation of the Convention;

(e) Facilitate the transfer of technologies, including biotechnology, to developing countries, consistent with the provisions of the Convention;

(f) Respect, preserve and maintain knowledge, innovations and practices of indigenous and local communities embodying traditional lifestyles, and encourage equitable sharing of the benefits arising from traditional knowledge so that those communities are adequately protected and rewarded, consistent with the provisions of the Convention on Biological Diversity and in accordance with the decisions of the Conference of the Parties;

(g) Complete rapidly the biosafety protocol under the Convention on Biological Diversity, on the understanding that the UNEP International Technical Guidelines for Safety in Biotechnology may be used as an interim mechanism during its development, and to complement it after its conclusion, including the recommendations on capacity-building related to biosafety;

(h) Stress the importance of the establishment of a clearing-house mechanism by Parties consistent with the provisions of the Convention;

(i) Recognize the role of women in the conservation of biological diversity and the sustainable use of biological resources;

(j) Provide the necessary support to integrate the conservation of biological diversity and the sustainable use of biological resources into national development plans;

(k) Promote international cooperation to develop and strengthen national capacity-building, including human resource development and institution-building;

(l) Provide incentive measures at the national, regional and international levels to promote the conservation and sustainable use of biological diversity, and to consider means to enhance developing countries' capabilities to compete in the emerging market for biological resources, while improving the functioning of that market.

Sustainable tourism

67. Tourism is now one of the world's largest industries and one of its fastest growing economic sectors. The expected growth in the tourism sector and the increasing reliance of many developing countries, including small island developing States, on this sector as a major employer and contributor to local, national, subregional and regional economies highlights the need for special attention to the relationship between environmental conservation and protection and sustainable tourism. In this regard, efforts of developing countries to broaden the traditional concept of tourism to include cultural and eco-tourism merit special consideration and the assistance of the international community, including the international financial institutions.

68. There is a need to consider further the importance of tourism in the context of Agenda 21. Tourism, like other sectors, uses resources and generates wastes, and creates environmental, cultural and social costs and benefits in the process. For sustainable patterns of consumption and production in the tourism sector, it is essential to strengthen national policy development and enhance capacity in the areas of physical planning, impact assessment, and the use of economic and regulatory instruments, as well as in the areas of information, education and marketing. A particular concern is the degradation of biodiversity and fragile ecosystems, such as coral reefs, mountains, coastal areas and wetlands.

69. Policy development and implementation should take place in cooperation with all interested parties, especially the private sector and local and indigenous communities. The Commission should develop an action-oriented international programme of

work on sustainable tourism, to be defined in cooperation with
the World Tourism Organization, UNCTAD, UNEP, the Conference
of the Parties to the Convention on Biological Diversity and other
relevant bodies.

70. The sustainable development of tourism is of importance for all
countries, in particular for small island developing States.
International cooperation is needed to facilitate tourism development in developing countries – including the development and
marketing of ecotourism, bearing in mind the importance of the
conservation policies required to secure long-term benefits from
development in this sector – in particular in small island developing States, in the context of the Programme of Action for the
Sustainable Development of Small Island Developing States.

Small island developing States

71. The international community reaffirms its commitment to the
implementation of the Programme of Action for the Sustainable
Development of Small Island Developing States. The Commission
on Sustainable Development carried out a mid-term review of
selected programme areas of the Programme of Action at its
fourth session, in 1996. In 1998, at its sixth session, the
Commission will undertake a review of all the outstanding
chapters and issues of the Programme of Action. A full and
comprehensive review of the Programme of Action, consistent
with the review of other United Nations global conferences is
scheduled for 1999. The Commission, at its fifth session, adopted
a decision on modalities for the full review of the Programme of
Action, including the holding of a two-day special session of the
General Assembly immediately preceding the fifty-fourth session
of the Assembly for an in-depth assessment and appraisal of the
implementation of the Programme of Action. The full implementation of the decision would represent a significant contribution
to achieving the objectives of the Global Conference for the
Sustainable Development of Small Island Developing States.

72. Considerable efforts are being made at the national and regional
levels to implement the Programme of Action. These efforts need
to be supplemented by effective financial support from the international community. External assistance for building the requisite
infrastructure and for national capacity-building, including
human and institutional capacity, and for facilitating access to
information on sustainable development practices and transfer of

environmentally sound technologies in accordance with paragraph 34.14 (b) of Agenda 21 is crucial for small island developing States to effectively attain the goals of the Programme of Action. To assist national capacity building, the small island developing States information network and small island developing States technical assistance programme should be operationalized as soon as possible, with support to existing regional and subregional institutions.

Natural disasters

73. Natural disasters have disproportionate consequences for developing countries, in particular small island developing States and countries with extremely fragile ecosystems. Programmes for sustainable development should give higher priority to implementation of the commitments made at the World Conference on Natural Disaster Reduction (Yokohama, Japan, 23–27 May 1994) (see A/CONF.172/9 and Add.1). There is a particular need for capacity-building for disaster planning and management and for the promotion and facilitation of the transfer of early-warning technologies to countries prone to disasters, in particular developing countries and countries with economies in transition.

74. Acknowledging that further work is needed throughout the world, there is a special need to provide developing countries with further assistance in:

    (a) Strengthening mechanisms and policies designed to reduce the effects of natural disasters, improve preparedness and integrate natural disaster considerations in development planning, through, inter alia, access to resources for disaster mitigation and preparedness, response and recovery;

    (b) Improving access to relevant technology and training in hazard and risk assessment and early warning systems, and in protection from environmental disasters, consistent with national, subregional and regional strategies;

    (c) Providing and facilitating technical, scientific and financial support for disaster preparedness and response in the context of the International Decade for Natural Disaster Reduction.

Major technological and other disasters with an adverse impact on the environment

75. Major technological and other disasters with an adverse impact

on the environment can be a substantial obstacle in the way of achieving the goals of sustainable development in many countries. The international community should intensify cooperation in the prevention and reduction of such disasters and in disaster relief and post-disaster rehabilitation in order to enhance the capabilities of affected countries to cope with such situations.

3. MEANS OF IMPLEMENTATION

Financial resources and mechanisms

76. Financial resources and mechanisms play a key role in the implementation of Agenda 21. In general, the financing for the implementation of Agenda 21 will come from a country's own public and private sectors. For developing countries, ODA is a main source of external funding, and substantial new and additional funding for sustainable development and implementation of Agenda 21 will be required. Hence, all financial commitments of Agenda 21, particularly those contained in chapter 33, and the provisions on new and additional resources that are both adequate and predictable need to be urgently fulfilled. Renewed efforts are essential to ensure that all sources of funding contribute to economic growth, social development and environmental protection in the context of sustainable development and the implementation of Agenda 21.

77. For developing countries, particularly those in Africa and the least developed countries, ODA remains a main source of external funding and is essential for the prompt and effective implementation of Agenda 21 and cannot generally be replaced by private capital flows. Developed countries should therefore fulfil the commitments undertaken to reach the accepted United Nations target of 0.7 per cent of GNP as soon as possible. In this context the present downward trend in the ratio of ODA to GNP causes concern. Intensified efforts should be made to reverse this trend, taking into account the need for improving the quality and effectiveness of ODA. In the spirit of global partnership, the underlying factors that have led to this decrease should be addressed by all countries. Strategies should be worked out for increasing donor support to aid programmes and revitalizing the commitments that donors made at UNCED. Some countries already meet or exceed the 0.7 per cent agreed target. Official financial flows to developing countries, particularly least devel-

oped countries, remain an essential element of the partnership embodied in Agenda 21. ODA plays a significant role, inter alia, in capacity-building, infrastructure, combating poverty and environmental protection in developing countries, and a crucial role in the least developed countries. ODA can play an important complementary and catalytic role in promoting economic growth and may, in some cases, play a catalytic role in encouraging private investment and, where appropriate, all aspects of country-driven capacity building and strengthening.

78. Funding by multilateral financial institutions through their concessional mechanisms is also essential to developing countries in their efforts to fully implement the sustainable development objectives contained in Agenda 21. Such institutions should continue to respond to the development needs and priorities of developing countries. Developed countries should urgently meet their commitments under the eleventh replenishment of the International Development Association (IDA).

79. Continued and full donor commitments to adequate, sustained and predictable funding for GEF operations is important for developing countries so that global environmental benefits can be further achieved. Donor countries are urged to engage in providing new and additional resources, with a view to equitable burden-sharing, through a satisfactory replenishment of GEF, which makes available grant and concessional funding designed to achieve global environmental benefits, thereby promoting sustainable development. Consideration should be given to further exploring the flexibility of the existing mandate of GEF in supporting activities to achieve global environmental benefits. With regard to the project cycle, further efforts should be made to continue streamlining the decision-making process in order to maintain an effective and efficient, as well as transparent, participatory and democratic framework. GEF, when acting as the operating entity of the financial mechanism of the United Nations Framework Convention on Climate Change and the Convention on Biological Diversity, should continue to operate in conformity with those Conventions and promote their implementation. The GEF implementing agencies, the United Nations Development Programme (UNDP), UNEP and the World Bank, should strengthen, as appropriate and in accordance with their respective mandates, their cooperation at all levels, including the field level.

80. The efficiency, effectiveness and impact of the operational activities of the United Nations system must be enhanced by, inter alia, a substantial increase in their funding on a predictable, continuous and assured basis, commensurate with the increasing needs of developing countries, as well as through the full implementation of resolutions 47/199 and 48/162. There is a need for a substantial increase in resources for operational activities for development on a predictable, continuous and assured basis, commensurate with the increasing needs of developing countries.

81. Private capital is a major tool of economic growth in a growing number of developing countries. Higher levels of foreign private investment should be mobilized given its mounting importance. To stimulate higher levels of private investment, Governments should aim at ensuring macroeconomic stability, open trade and investment policies, and well-functioning legal and financial systems. Further studies should be undertaken, including studies on the design of an appropriate environment, at both the national and international levels, for facilitating foreign private investment, in particular foreign direct investment (FDI) flows to developing countries, and enhancing its contribution to sustainable development. To ensure that such investments are supportive of sustainable development objectives, it is essential that the national Governments of both investor and recipient countries provide appropriate regulatory frameworks and incentives for private investment. Therefore further work should be undertaken on the design of appropriate policies and measures aimed at promoting long-term investment flows to developing countries in activities which increase their productive capability, and reducing the volatility of these flows. ODA donors and multilateral development banks are encouraged to strengthen their commitments to supporting investment in developing countries in a manner that jointly promotes economic growth, social development and environmental protection.

82. The external debt problem continues to hamper the efforts of developing countries to achieve sustainable development. To resolve the remaining debt problems of the heavily indebted poor countries, creditor and debtor countries and international financial institutions should continue their efforts to find effective, equitable, development-oriented and durable solutions to the debt problem, including debt relief in the form of debt

rescheduling, debt reduction, debt swaps and, as appropriate, debt cancellation, as well as grants and concessional flows that will help restore creditworthiness. The joint World Bank/International Monetary Fund (IMF) Heavily Indebted Poor Countries (HIPC) Debt Initiative supported by the Paris Club creditor countries is an important development to reduce the multilateral debt problem. Implementation of the HIPC Debt Initiative requires additional financial resources from both bilateral and multilateral creditors without affecting the support required for the development activities of developing countries.

83. There is a need for a fuller understanding of the impact of indebtedness on the pursuit of sustainable development by developing countries. To this end, the United Nations Secretariat, the World Bank and IMF are invited to collaborate with UNCTAD in further considering the interrelationship between indebtedness and sustainable development for developing countries.

84. While international cooperation is very important in assisting developing countries in their development efforts, in general financing for the implementation of Agenda 21 will come from countries' own public and private sectors. Policies for promoting domestic resource mobilization, including credit, could include sound macroeconomic reforms, including fiscal and monetary policy reforms, review and reform of existing subsidies, and the promotion of personal savings and access to credit, especially micro-credit, in particular for women. Such policies should be decided by each country, taking into account its own characteristics and capabilities and different levels of development, especially as reflected in national sustainable development strategies, where they exist.

85. There is a need for making existing subsidies more transparent in order to increase public awareness of their actual economic, social and environmental impact and to reform or, where pertinent, remove them. Further national and international research in that area should be promoted in order to assist Governments in identifying and considering phasing-out subsidies that have market distorting, and socially and environmentally damaging impacts. Subsidy reductions should take full account of the specific conditions and the different levels of development of individual countries and should consider potentially regressive impacts, particularly on developing countries. In addition, it would be desirable to use international cooperation and coordi-

nation to promote the reduction of subsidies where these have important implications for competitiveness.

86. In order to reduce the barriers to the expanded use of economic instruments, Governments and international organizations should collect and share information on the use of economic instruments and introduce pilot schemes that would, inter alia, demonstrate how to make the best use of such instruments while avoiding adverse effects on competitiveness and terms of trade of all countries, particularly developing countries, and on marginalized and vulnerable sectors of society. When introducing economic instruments that raise the cost of economic activities for households and small and medium-sized enterprises (SMEs), Governments should consider gradual phase-ins, public education programmes and targeted technical assistance as strategies for reducing distributional impacts. Various studies and practical experiences in a number of countries, in particular developed countries, indicate that the appropriate use of relevant economic instruments may help generate positive possibilities for shifting consumer and producer behaviour to more sustainable directions in those countries. There is, however, a need to conduct further studies and test practical experiences in more countries, taking into account country-specific conditions, and particularly the acceptability, legitimacy, equity, efficiency and effectiveness of such economic instruments.

87. Innovative financial mechanisms are currently under discussion in international and national forums but have not yet fully evolved conceptually. The Secretary-General is to submit a report concerning innovative financing mechanisms to the Economic and Social Council at its substantive session of 1997. In view of the widespread interest in those mechanisms, appropriate organizations, including UNCTAD, the World Bank and IMF, are invited to consider conducting forward-looking studies of concerted action on such mechanisms and to share them with the Commission on Sustainable Development, other relevant intergovernmental organizations and non-governmental organizations. In this regard, innovative funding should complement ODA, not replace it. New initiatives for cooperative implementation of environment and development objectives under mutually beneficial incentive structures should be further explored.

Transfer of environmentally sound technologies

88. The availability of scientific and technological information and access to and transfer of environmentally sound technologies are essential requirements for sustainable development. There is an urgent need for developing countries to acquire greater access to environmentally sound technologies if they are to meet the obligations agreed at UNCED and in the relevant international conventions. The ability of developing countries to participate in, benefit from and contribute to rapid advances in science and technology can significantly influence their development. This calls for the urgent fulfillment of all the UNCED commitments concerning concrete measures for the transfer of environmentally sound technologies to developing countries. The international community should promote, facilitate and finance, as appropriate, access to and transfer of environmentally sound technologies and corresponding know-how, in particular to developing countries, on favourable terms, including concessional and preferential terms, as mutually agreed, taking into account the need to protect intellectual property rights as well as the special needs of developing countries for the implementation of Agenda 21. Current forms of cooperation involving the public and private sectors of developing and developed countries should be built upon and expanded. In this context, it is important to identify barriers and restrictions to the transfer of publicly and privately owned environmentally sound technologies, with a view to reducing such constraints while creating specific incentives, fiscal and otherwise, for the transfer of such technologies. The progress on the fulfillment of all provisions as contained in chapter 34 of Agenda 21 should be regularly reviewed as part of the multi-year work programme of the Commission on Sustainable Development.

89. Technology transfer and the development of the human and institutional capacity to adapt, absorb and disseminate technologies, as well as to generate technical knowledge and innovations, are part of the same process and must be given equal importance. Governments have an important role to play in providing, inter alia, research and development institutions with incentives to promote and to contribute to the development of institutional and human capacities.

90. Much of the most advanced environmentally sound technology is developed and held by the private sector. Creation of an enabling

environment, on the part of both developed and developing countries, including supportive economic and fiscal measures, as well as a practical system of environmental regulations and compliance mechanisms, can help to stimulate private sector investment in and transfer of environmentally sound technology to developing countries. New ways of financial intermediation for the financing of environmentally sound technologies, such as "green credit lines", should be examined. Further efforts should be made by Governments and international development institutions to facilitate the transfer of privately owned technology on concessional terms, as mutually agreed, to developing countries, especially least developed countries.

91. A proportion of technology is held or owned by Governments and public institutions or results from publicly funded research and development activities. The Government's control and influence over the technological knowledge produced in publicly funded research and development institutions open up the potential for the generation of publicly owned technologies that could be made accessible to developing countries, and could be an important means for Governments to catalyze private sector technology transfer. Proposals for further study of the options with respect to those technologies and publicly funded research and development activities are to be welcomed.

92. Governments should create a legal and policy framework that is conducive to technology-related private sector investments and long-term sustainable development objectives. Governments and international development institutions should continue to play a key role in establishing public-private partnerships, within and between developed and developing countries and countries with economies in transition. Such partnerships are essential for linking the advantages of the private sector – access to finance and technology, managerial efficiency, entrepreneurial experience and engineering expertise – with the capacity of Governments to create a policy environment that is conducive to technology-related private sector investments and long-term sustainable development objectives.

93. The creation of centres for the transfer of technology at various levels, including the regional level, could greatly contribute to achieving the objective of transfer of environmentally sound technologies to developing countries. For this purpose, existing United Nations bodies and mechanisms, including, as appropri-

ate, TCDC, ECDC, the Commission on Science and Technology for Development, UNCTAD, the United Nations Industrial Development Organization (UNIDO), UNEP and the regional commissions, should cooperate.

94. Governments and international development institutions can also play an important role in bringing together companies from developed and developing countries and countries with economies in transition so that they can create sustainable and mutually beneficial business linkages. Incentives should be given to stimulate the building of joint ventures between small and medium-sized enterprises of developed and developing countries and countries with economies in transition, and cleaner production programmes in public and private companies should be supported.

95. Governments of developing countries should take appropriate measures to strengthen South–South cooperation for technology transfer and capacity-building. Such measures could include the networking of existing national information systems and sources on environmentally sound technologies, and the networking of national cleaner production centres, as well as the establishment of sector-specific regional centres for technology transfer and capacity-building. Interested donor countries and international organizations should further assist developing countries in those efforts through, inter alia, supporting trilateral arrangements and contributing to the United Nations Voluntary Trust Fund for South–South Cooperation.

96. Attention must also be given to technology needs assessment as a tool for Governments in identifying a portfolio for technology transfer projects and capacity-building activities to be undertaken to facilitate and accelerate the development, adoption and dissemination of environmentally sound technologies in particular sectors of the national economy. It is also important for Governments to promote the integration of environmental technology assessment with technology needs assessment as an important tool for evaluating environmentally sound technologies and the organizational, managerial and human resource systems related to the proper use of those technologies.

97. There is a need to further explore and enhance the potential of global electronic information and telecommunication networks. This would enable countries to choose among the available technological options that are most appropriate to their needs. In

this respect, the international community should assist developing countries to enhance their capacities.

Capacity-building

98. Renewed commitment and support from the international community is essential to support national efforts for capacity-building in developing countries and countries with economies in transition.

99. The United Nations Development Programme, inter alia, through its Capacity 21 programme, should give priority attention to building capacity for the elaboration of sustainable development strategies based on participatory approaches. In this context, developing countries should be assisted, particularly in the areas of the design, implementation and evaluation of programmes and projects.

100. Capacity-building efforts should pay particular attention to the needs of women in order to ensure that their skills and experience are fully used in decision-making at all levels. The special needs, culture, traditions and expertise of indigenous people must be recognized. International financial institutions should continue to give high priority to funding capacity-building for sustainable development in developing countries and countries with economies in transition. Special attention should also be given to strengthening the ability of developing countries to absorb and generate technologies. International cooperation needs to be strengthened to promote the endogenous capacity of developing countries to utilize scientific and technological developments from abroad and to adapt them to local conditions. The role of the private sector in capacity-building should be further promoted and enhanced. South–South cooperation in capacity-building should be further supported through "triangular" cooperative arrangements. Both developed and developing countries, in cooperation with relevant international institutions, need to strengthen their efforts to develop and implement strategies for more effective sharing of environmental expertise and data.

Science

101. Public and private investment in science, education and training, and research and development should be increased significantly, with emphasis on the need to ensure equal access to opportunities for girls and women.

102.International consensus-building is facilitated by the availability
of authoritative scientific evidence. There is a need for further
scientific cooperation, especially across disciplines, in order to
verify and strengthen scientific evidence and make it accessible to
developing countries. This evidence is important for assessing
environmental conditions and changes. Steps should also be
taken by Governments, academia, and scientific institutions to
improve access to scientific information related to the environment and sustainable development. Promotion of existing
regional and global networks may be useful for this purpose.

103.Increasing efforts to build and strengthen scientific and technological capacity in developing countries is an extremely important
objective. Multilateral and bilateral donor agencies and
Governments, as well as specific funding mechanisms, should
continue to enhance their support for developing countries.
Attention should also be given to countries with economies in
transition.

104.The international community should also actively collaborate to
promote innovations in information and communication
technologies for the purpose of reducing environmental impacts,
inter alia, by taking user-needs based approaches to technology
transfer and cooperation.

Education and awareness

105.Education increases human welfare, and is a decisive factor in
enabling people to become productive and responsible members
of society. A fundamental prerequisite for sustainable development is an adequately financed and effective educational system
at all levels, particularly the primary and secondary levels, that is
accessible to all and that augments both human capacity and
well-being. The core themes of education for sustainability
include lifelong learning, interdisciplinary education, partnerships, multicultural education and empowerment. Priority should
be given to ensuring women's and girls' full and equal access to
all levels of education and training. Special attention should also
be paid to the training of teachers, youth leaders and other
educators. Education should also be seen as a means of empowering youth and vulnerable and marginalized groups, including
those in rural areas, through intergenerational partnerships and
peer education. Even in countries with strong education systems,
there is a need to reorient education, awareness and training to

increase widespread public understanding, critical analysis and
support for sustainable development. Education for a sustainable
future should engage a wide spectrum of institutions and sectors,
including but not limited to business/industry, international
organizations, youth, professional organizations, non-governmental organizations, higher education, government, educators and
foundations, to address the concepts and issues of sustainable
development, as embodied throughout Agenda 21, and should
include the preparation of sustainable development education
plans and programmes, as emphasized in the Commission's work
programme on the subject adopted in 1996.[28] The concept of
education for a sustainable future will be further developed by
the United Nations Educational, Scientific and Cultural
Organization, in cooperation with others.

106. It is necessary to support and strengthen universities and other
academic centres in promoting cooperation among them, particularly cooperation between those of developing countries and
those of developed countries.

International legal instruments and the Rio Declaration on Environment and Development

107. The implementation and application of the principles contained
in the Rio Declaration on Environment and Development should
be the subject of regular assessment and reporting to the
Commission on Sustainable Development by the Secretariat in
collaboration with UNEP, in particular.

108. Access to information and public participation in decision-making
are fundamental to sustainable development. Further efforts are
required to promote, in the light of country-specific conditions,
the integration of environment and development policies,
through appropriate legal and regulatory policies, instruments
and enforcement mechanisms at the national, state, provincial
and local levels. At the national level, each individual shall have
appropriate access to information concerning the environment
that is held by public authorities, including information on
hazardous materials and activities in the communities, and the
opportunity to participate in decision-making processes.
Governments and legislators, with the support, where appropriate, of competent international organizations, should establish
judicial and administrative procedures for legal redress and
remedy of actions affecting environment and development that

may be unlawful or infringe on rights under the law, and should provide access to individuals, groups and organizations with a recognized legal interest. Access should be provided to effective judicial and administrative channels for affected individuals and groups to ensure that all authorities, both national and local, and other civil organizations remain accountable for their actions in accordance with their obligations, at the appropriate levels for the country concerned, taking into account the judicial and administrative systems of the country concerned.

109. Taking into account the provisions of chapter 39, particularly paragraph 39.1, of Agenda 21, it is necessary to continue the progressive development and, as and when appropriate, codification of international law related to sustainable development. Relevant bodies, where such tasks are being undertaken, should cooperate and coordinate in this regard.

110. Implementation of and compliance with commitments made under international treaties and other instruments in the field of environment remains a priority. Implementation can be promoted by secure, sustained and predictable financial support, sufficient institutional capacity, human resources and adequate access to technology. Cooperation on implementation between States on mutually agreed terms may help reduce potential sources of conflict between States. In this context, States should further study and consider methods to broaden and make more effective the range of techniques available at present, taking into account relevant experience under existing agreements and, where appropriate, modalities for dispute avoidance and settlement, in accordance with the Charter of the United Nations. It is also important to further improve reporting and data-collection systems and to further develop appropriate compliance mechanisms and procedures, on a mutually agreed basis, to help and encourage States to fulfil all their obligations, including means of implementation, under multilateral environmental agreements. Developing countries should be assisted to develop these tools according to country-specific conditions.

Information and tools to measure progress

111. The further development of cost-effective tools to collect and disseminate information for decision makers at all levels through strengthened data collection including, as appropriate, gender disaggregated data, including information that makes visible the

unremunerated work of women for use in programme planning and implementation, compilation and analysis is urgently needed. In this context, emphasis will be placed on support of national and international scientific and technological data centres with appropriate electronic communication links between these centres.

112. A supportive environment needs to be established to enhance national capacities and capabilities for information collection, processing and dissemination, especially in developing countries, to facilitate public access to information on global environmental issues through appropriate means including high-tech information and communication infrastructure related to the global environment, in the light of country-specific conditions, using, where available, such tools as geographic information systems and video transmission technology, including global mapping. In this regard, international cooperation is essential.

113. Environmental Impact Assessments (EIAs) are an important national tool for sustainable development. In accordance with Principle 17 of the Rio Declaration, EIAs shall be undertaken for proposed activities that are likely to have a significant adverse impact on the environment and are subject to a decision of a competent national authority, and, where appropriate, they shall be made available early in the project cycle.

114. The Commission's work programme on indicators of sustainable development should result in a practicable and agreed set of indicators, suited to country-specific conditions, including a limited number of aggregated indicators, to be used at the national level, on a voluntary basis, by the year 2000. Such indicators of sustainable development, including, where appropriate, and subject to nationally specific conditions, sector-specific ones, should play an important role in monitoring progress towards sustainable development at the national level and in facilitating national reporting, as appropriate.

115. National reports on the implementation of Agenda 21 have proved to be a valuable means of sharing information at the international and regional levels and, even more important, of providing a focus for the coordination of issues related to sustainable development at the national level within individual countries. National reporting should continue. (See also paragraph 133 (b) and (c).)

D. INTERNATIONAL INSTITUTIONAL ARRANGEMENTS

116.The achievement of sustainable development requires continued support from international institutions. The institutional framework outlined in chapter 38 of Agenda 21 and determined by the General Assembly in its resolution 47/191 and other relevant resolutions, including the specific functions and roles of various organs, organizations and programmes within and outside the United Nations system, will continue to be fully relevant in the period after the special session of the General Assembly. In the light of the ongoing discussions on reform within the United Nations, international institutional arrangements in the area of sustainable development are intended to contribute to the goal of strengthening the entire United Nations system. In this context, the strengthening of the institutions for sustainable development, as well as the achievement of the goals and objectives set out below would be particularly important.

1. GREATER COHERENCE IN VARIOUS INTERGOVERNMENTAL ORGANIZATIONS AND PROCESSES

117.Given the increasing number of decision-making bodies concerned with various aspects of sustainable development, including international conventions, there is an ever greater need for better policy coordination at the intergovernmental level, as well as for continued and more concerted efforts to enhance collaboration among the secretariats of those decision-making bodies. Under the guidance of the General Assembly, the Economic and Social Council should play a strengthened role in coordinating the activities of the United Nations system in the economic, social and related fields.

118.The conferences of the parties to conventions signed at the Rio Conference or as a result of it, as well as other conventions related to sustainable development, should cooperate in exploring ways and means of collaborating in their work to advance the effective implementation of the conventions. There is also a need for environmental conventions to continue to pursue sustainable development objectives consistent with their provisions and be fully responsive to Agenda 21. To this end, inter alia, the conferences of the parties or governing bodies of the conventions signed at the Rio Conference, or as a result of it and of other relevant conventions and agreements should, if appropriate, give consideration to the co-location of secretariats, to improving the scheduling of meetings, to integrating national reporting requirements, to improving the balance between sessions of the conferences of the parties and sessions of their subsidiary bodies, and to encouraging and facilitating the participation of Governments in those sessions, at an appropriate level.

119. Institutional arrangements for the convention secretariats should provide effective support and efficient services, while ensuring that in order for them to be efficient, at their respective locations, appropriate autonomy is necessary. At the international and national levels there is a need for, inter alia, better scientific assessment of ecological linkages between the conventions; identification of programmes that have multiple benefits; and enhanced public awareness-raising for the conventions. Such tasks should be undertaken by UNEP in accordance with the relevant decisions of its Governing Council and in full cooperation with the conferences of the parties to and governing bodies of relevant conventions. Efforts of convention secretariats, in response to requests from the respective conferences of the parties, to explore, where appropriate, modalities for suitable liaison arrangements in Geneva and/or New York for the purpose of enhancing linkages with delegations and organizations at those United Nations centres are welcomed and fully supported.

120. It is necessary to strengthen the ACC Inter-Agency Committee on Sustainable Development and its system of task managers, with a view to further enhancing system-wide intersectoral cooperation and coordination for the implementation of Agenda 21 and for the promotion of coordinated follow-up to the major United Nations conferences in the area of sustainable development.

121. The Commission on Sustainable Development should promote increased regional implementation of Agenda 21 in cooperation with relevant regional and subregional organizations and the United Nations regional commissions, in accordance with the results of their priority-setting efforts, with a view to enhancing the role such bodies play in the achievement of sustainable development objectives agreed at the international level. The regional commissions could provide appropriate support, consistent with their work programmes, to regional meetings of experts related to the implementation of Agenda 21.

2. ROLE OF RELEVANT ORGANIZATIONS AND INSTITUTIONS OF THE UNITED NATIONS SYSTEM

122. In order to facilitate the national implementation of Agenda 21, all organizations and programmes of the United Nations system, within their respective areas of expertise and mandates, should strengthen, individually and jointly, the support for national efforts to implement Agenda 21 and make their efforts and actions consistent with national plans, policies and priorities of member States. Coordination of United Nations activities at the field level should be further enhanced through the resident coordinator system in full consultation with national Governments.

123. The role of UNEP, as the principal United Nations body in the field of environment, should be further enhanced. Taking into account its catalytic role, and in conformity with Agenda 21 and the Nairobi Declaration on the Role and Mandate of the United Nations Environment Programme, adopted on 7 February 1997,[29] UNEP is to be the leading global environmental authority that sets the global environmental agenda, promotes the coherent implementation of the environmental dimension of sustainable development within the United Nations system, and serves as an authoritative advocate for the global environment. In this context, the Governing Council decision of 4 April 1997 on governance and other related Governing Council decisions are relevant. The role of UNEP in the further development of international environmental law should be strengthened, including the development of coherent interlinkages among relevant environmental conventions in cooperation with their respective conferences of the parties or governing bodies. In performing its functions related to the conventions signed at the Rio Conference or as a result of it and other relevant conventions, UNEP should strive to promote the effective implementation of those conventions in a manner consistent with the provisions of the conventions and the decisions of the conferences of the parties.

124. UNEP, in the performance of its role, should focus on environmental issues, taking into account the development perspective. A revitalized UNEP should be supported by adequate, stable and predictable funding. UNEP should continue providing effective support to the Commission on Sustainable Development, inter alia, in the form of scientific, technical and policy information, analysis and advice on global environmental issues.

125.UNDP should continue to strengthen its contribution to programmes in sustainable development and the implementation of Agenda 21 at all levels particularly in the area of promoting capacity-building (including through its Capacity 21 programme) in cooperation with other organizations, as well as in the field of poverty eradication.

126.UNCTAD, in accordance with General Assembly resolution 51/167 and relevant decisions of the Trade and Development Board on the work programme, should continue to play a key role in the implementation of Agenda 21 through the integrated examination of linkages among trade, investment, technology, finance and sustainable development.

127.The WTO Committee on Trade and Environment, UNCTAD and UNEP should advance their coordinated work on trade and environment, involving other appropriate international and regional organizations in their cooperation and coordination. In coordination with WTO, UNCTAD and UNEP should continue to support efforts to promote the integration of trade, environment and development. The Commission on Sustainable Development should continue to play its important role in the deliberations on trade and environment to facilitate the integrated consideration of all factors relevant for achieving sustainable development.

128.Implementation of the commitment of the international financial institutions to sustainable development should continue to be strengthened. The World Bank has a significant role to play, bearing in mind its expertise and the overall volume of resources that it commands.

129.Operationalization of the global mechanism of the United Nations Convention to Combat Desertification in Those Countries Experiencing Serious Drought and/or Desertification, Particularly in Africa is also essential.

3. FUTURE ROLE AND PROGRAMME OF WORK OF THE COMMISSION ON SUSTAINABLE DEVELOPMENT

130.The Commission on Sustainable Development, within its mandate, as specified in General Assembly resolution 47/191, will continue to provide a central forum for reviewing progress and for urging further implementation of Agenda 21 and other commitments made at UNCED or as a result of it, for conducting high-level policy debate aimed at consensus-building on sustainable development and for catalyzing action and long-term commitment to sustainable development at all levels. It should continue to undertake these tasks in complementing and providing interlinkages to the work of other United Nations organs, organizations and bodies acting in the field of sustainable development. The Commission has a role to play in assessing the challenges of globalization as they relate to sustainable development. The Commission should perform its functions in coordination with other subsidiary bodies of the Economic and Social Council and other related organizations and institutions, including making recommendations, within its mandate, to the Economic and Social Council, bearing in mind the interrelated outcomes of recent United Nations conferences.

131. The Commission should focus on issues that are crucial to achieving the goals of sustainable development. It should promote policies that integrate economic, social and environmental dimensions of sustainability and should provide for integrated consideration of linkages, both among sectors and between sectoral and cross-sectoral aspects of Agenda 21. In this connection, the Commission should carry out its work in such a manner as to avoid unnecessary duplication and repetition of work undertaken by other relevant forums.

132. In the light of the above, it is recommended that the Commission on Sustainable Development adopt the multi-year programme of work for the period 1998–2002 contained in the annex below.

4. METHODS OF WORK OF THE COMMISSION ON SUSTAINABLE DEVELOPMENT

133. Based on the experience gained during the period 1993–1997, the Commission, under the guidance of the Economic and Social Council, should:

(a) Make concerted efforts to attract greater involvement in its work of ministers and high-level national policy makers responsible for specific economic and social sectors, who, in particular, are encouraged to participate in the annual high-level segments of the Commission together with the ministers and policy makers responsible for environment and development. The high-level segments of the Commission should become more interactive, and should focus on the priority issues being considered at a particular session. The Bureau of the Commission should conduct timely and open-ended consultations with the view to improving the organization of the work of the high-level segments;

(b) Continue to provide a forum for the exchange of national experience and best practices in the area of sustainable development, including through voluntary national communications or reports. Consideration should be given to the results of ongoing work aimed at streamlining requests for national information and reporting and the results of the "pilot phase" on indicators of sustainable development. In this context, the Commission should consider more effective modalities for the further implementation of commitments made in Agenda 21, with an appropriate emphasis on means of implementation. Countries may wish to submit to the Commission, on a voluntary basis, information regarding their efforts to incorporate the relevant recommendations of other United Nations conferences in national sustainable development strategies;

(c) The Commission should take into account regional developments related to the implementation of the outcomes of UNCED. It should provide a forum for the exchange of experience on regional and subregional initiatives and regional collaboration for sustainable development. This could include the promotion of the voluntary regional exchange of national experience in the implementation of Agenda 21 and, inter alia, the possible development of modalities for reviews by and among those countries that voluntarily agree to do so, within regions. In this context, the Commission should encourage the availability of funding for the implementation of initiatives related to such reviews;

(d) Establish closer interaction with international financial, development and trade institutions, as well as with other relevant bodies within and outside the United Nations system, including the World Bank, GEF, UNDP, WTO, UNCTAD and UNEP, which, in turn, are invited to take full account of the results of policy deliberations in the Commission and to integrate them in their own work programmes and activities;

(e) Strengthen its interaction with representatives of major groups including through greater and better use of focused dialogue sessions and round tables. These groups are important resources in operationalizing, managing and promoting sustainable development and contribute to the implementation of Agenda 21. The major groups are encouraged to adopt arrangements for coordination and interaction in providing inputs to the Commission. Taking into account the Commission's programme of work, this could include inputs from:

(i) The scientific community and research institutions on greater understanding of the interactions between human activity and natural ecosystems and on how to manage global systems sustainably;

(ii) Women, children and youth, indigenous people and their communities, non-governmental organizations, local authorities, workers and their trade unions and farmers on the elaboration, promotion and sharing of effective strategies, policies, practices and processes to promote sustainable development;

(iii) Business and industry groups in the elaboration, promotion and sharing of sustainable development practices and their promotion of corporate responsibility and accountability;

(f) Organize the implementation of its next multi-year programme of work in the most effective and productive way, including through shortening its annual meeting to two weeks. The inter-sessional ad hoc working groups should help to focus the Commission's sessions by identifying key elements to be discussed and important problems to be addressed within specific items of the Commission's programme of work. Government hosted and funded expert meetings will continue to provide inputs to the work of the Commission.

134.The Secretary-General is invited to review the functioning of the High-Level Advisory Board on Sustainable Development and present proposals on ways to promote more direct interaction between the Board and the Commission, with a view to ensuring that the Board contributes to the deliberations on specific themes considered by the Commission in accordance with its programme of work.

135.The work of the Committee on New and Renewable Sources of Energy and on Energy for Development and the Committee on Natural Resources should be more compatible with and supportive of the programme of work of the Commission. The Economic and Social Council, in carrying out its functions related to the implementation of General Assembly resolution 50/227, should consider, at its substantive session of 1997, the most effective means of bringing this about.

136. Arrangements for the election of the Bureau should be changed in order to allow the same Bureau to provide guidance for the preparation for and lead work during the annual sessions of the Commission. The Commission would benefit from such a change, and the Economic and Social Council should take the necessary action at its substantive session of 1997 to ensure that these new arrangements take effect.

137. The next comprehensive review of progress achieved in the implementation of Agenda 21 by the General Assembly will take place in the year 2002. The modalities of this review should be determined at a later stage.

NOTES

1 United Nations Environment Programme, Global Environment Outlook (Oxford, Oxford University Press, 1997).

2 United Nations Environment Programme, Convention on Biological Diversity (Environmental Law and Institution Programme Activity Centre), June 1992.

3 Report of the Global Conference on the Sustainable Development of Small Island Developing States, Bridgetown, Barbados, 25 April–6 June 1994 (United Nations publication, Sales No. E.94.I.18 and corrigenda), chap. I, resolution 1, annex II.

4 United Nations Convention on the Law of the Sea with Index and Final Act of the Third United Nations Conference on the Law of the Sea (United Nations publication, Sales No. E.83.V.5).

5 Report of the United Nations Conference on Environment and Development, Rio de Janeiro, 3–14 June 1992, vol. I, Resolutions Adopted by the Conference (United Nations publication, Sales No. E.93.I.8 and corrigendum), resolution 1, annex I.

6 Ibid., resolution 1, annex III.

7 Report of the World Summit for Social Development, Copenhagen, 6–12 March 1995 (United Nations publication, Sales No. E.96.IV.8), chap. I, resolution 1, annex I.

8 Ibid., resolution 1, annex II.

9 Report of the Fourth World Conference on Women, Beijing, 4–15 September 1995 (A/CONF.177/20 and Add.1), chap. I, resolution 1, annex II. All references to the platforms or programmes for action of major conferences in this report should be considered in a manner consistent with their reports.

10 See, inter alia, A shared vision – conclusions from the chairperson of the Brasilia Workshop on sustainable production and consumption patterns and policies, held from 25–28 November 1996, Note of the Secretary-General (E/CN.17/1997/19), Appendix.

11 Results of the Uruguay Round of Multilateral Trade Negotiations: The Legal Texts (Geneva, GATT secretariat, 1994).

12 Adopted by the WTO Ministerial Meeting held at Singapore in December 1996.

13 Report of the International Conference on Population and Development, Cairo, 5–13 September 1994 (United Nations publication, Sales No. E.95.XIII.18). All references to the platforms or programmes for action of major conferences in this report should be considered in a manner consistent with their reports.

14 See Report of the International Conference on Primary Health Care, Alma-Ata, USSR, 6–12 September 1978 (Geneva, World Health Organization, 1978).

15 Report of the United Nations Conference on Human Settlements (Habitat II), Istanbul, 3–14 June 1996 (A/CONF.165/14), chap. I, resolution 1, annexes I and II.

16 Official Records of the Economic and Social Council, 1996, Supplement No. 8 (E/1996/28), chap. I, sect. C, decision 4/15, para. 45.

17 Ibid., 1997, Supplement No. 9 (E/1997/29).

18 See Legal Instruments Embodying the Results of the Uruguay Round of Multilateral Trade Negotiations, done at Marrakesh on 15 April 1994 (GATT secretariat publication, Sales No. GATT/1994–7).

19 Berlin Mandate: review of the adequacy of article 4, paragraph 2 (a) and (b) of the United Nations Framework Convention on Climate Change, including proposals related to a protocol and decisions on follow-up (FCCC/CP/1995/7/Add.1, sect. I, decision 1/CP.1).

20 Report of the Conference of the Parties to the United Nations Framework Convention on Climate Change on its second session, Geneva, 8–19 July 1996 (FCCC/CP/1996/15/Add.1), annex.

21 Montreal Protocol on Substances that Deplete the Ozone Layer, International Legal Materials, vol. 26, No. 6 (November 1987), p. 1550.

22 Basel Convention on the Control of Transboundary Movements of Hazardous Wastes and Their Disposal (UNEP/WG/190/4) (United Nations, Treaty Series, vol. 1673, No. 28911, forthcoming).

23 Bamako Convention on the Ban of the Import into Africa of All Forms of Hazardous Wastes and the Control of their Transboundary Movements within Africa, International Legal Materials, vol. 30, No. 3 (May 1991), p. 775, and vol. 32, No. 1 (January 1992), p. 164.

24 Where management appears in the section on radioactive wastes, it is defined as handling, treatment, storage, transportation, including transboundary movement, and final disposal of such wastes.

25 Report of the third meeting of the Conference of the Parties to the Convention on Biological Diversity (UNEP/CBD/COP/3/38), annex II.

26 Report of the World Food Summit, Rome, 13–17 November 1996, Part One (WFS 96/REP) (Rome, Food and Agriculture Organization of the United Nations, 1997), appendix.

27　Report of the third meeting of the Conference of the Parties to the Convention on Biological Diversity (UNEP/CBD/COP/3/38), annex II, decision III/12.

28　See Official Records of the Economic and Social Council, 1996, Supplement No. 8 (E/1996/28), chap. I, sect. C, decision 4/11.

29　Decision 19/1 of the Governing Council of the United Nations Environment Programme; reproduced in document A/S–19/5, annex, sect. I.

ANNEX

MULTI-YEAR PROGRAMME OF WORK FOR THE COMMISSION ON SUSTAINABLE DEVELOPMENT 1998–2002

1998 session:

Overriding issues: poverty/consumption and production patterns
Sectoral theme: STRATEGIC APPROACHES TO FRESHWATER MANAGEMENT
Review of outstanding chapters of the Programme of Action for the Sustainable Development of Small Island Developing States[a]
Main issues for an integrated discussion under the above theme: Agenda 21, chapters 2–8, 10–15, 18–21, 23–34, 36, 37, 40.
Cross-sectoral theme: TRANSFER OF TECHNOLOGY/ CAPACITY-BUILDING/EDUCATION/SCIENCE/ AWARENESS-RAISING
Main issues for an integrated discussion under the above theme: Agenda 21, chapters 2–4, 6, 16, 23–37, 40.
Economic sector/major group: INDUSTRY
Main issues for an integrated discussion under the above theme: Agenda 21, chapters 4, 6, 9, 16, 17, 19 21, 23 35, 40.

1999 session:

Overriding issues: poverty/consumption and production patterns
Comprehensive review of the Programme of Action for the Sustainable Development of Small Island Developing States
Sectoral theme: OCEANS AND SEAS
Main issues for an integrated discussion under the above theme: Agenda 21, chapters 5–7, 9, 15, 17, 19–32, 34–36, 39–40.
Cross-sectoral theme: CONSUMPTION AND PRODUCTION PATTERNS
Main issues for an integrated discussion under the above theme: Agenda 21, chapters 2–10, 14, 18–32, 34–36, 40.
Economic sector/Major group: TOURISM

Main issues for an integrated discussion under the above theme:
Agenda 21, chapters 2–7, 13, 15, 17, 23–33, 36.

2000 session:

Overriding issues: poverty/consumption and production patterns
Sectoral theme: INTEGRATED PLANNING AND MANAGEMENT OF
LAND RESOURCES
Main issues for an integrated discussion under the above theme:
Agenda 21, chapters 2–8, 10–37, 40.
Cross-sectoral theme: FINANCIAL RESOURCES/ TRADE AND
INVESTMENT/ECONOMIC GROWTH
Main issues for an integrated discussion under the above theme:
Agenda 21, chapters 2–4, 23–33, 36–38, 40.
Economic sector/major group: AGRICULTURE[b]
Day of Indigenous People
Main issues for an integrated discussion under the above theme:
Agenda 21, chapters 2–7, 10–16, 18–21, 23–34, 37, 40.

2001 session:

Overriding issues: poverty/consumption and production patterns
Sectoral theme: ATMOSPHERE; ENERGY
Main issues for an integrated discussion under the above theme:
Agenda 21, chapters 4, 6–9, 11–14, 17, 23–37, 39–40.
Cross-sectoral theme: INFORMATION FOR DECISION-MAKING AND
PARTICIPATION;
INTERNATIONAL COOPERATION FOR AN ENABLING ENVIRON-
MENT
Main issues for an integrated discussion under the above theme:
Agenda 21, chapters 2, 4, 6, 8, 23–36, 38–40.
Economic sector/major group: ENERGY; TRANSPORT
Main issues for an integrated discussion under the above theme:
Agenda 21, chapters 2–5, 8, 9, 20, 23–37, 40.

2002 session:

Comprehensive review

NOTES

a   Review to include those chapters of the SIDS Programme of Action not
     covered in the in-depth review carried out by the fourth session of the
     CSD.
b   Including forestry.

Index

Note: Acronyms, where not shown in the index, can be found on page vii at the front of the book.

Printed in the United States
by Baker & Taylor Publisher Services